Errata and Corrigenda no. 8 : Ja   en entered.
                        no. 9 : Ja

MASTER INDEX
To Materials and Properties

THERMOPHYSICAL PROPERTIES OF MATTER
The TPRC Data Series

A Comprehensive Compilation of Data by the
Thermophysical Properties Research Center (TPRC), Purdue University

Y. S. Touloukian, Series Editor
C. Y. Ho, Series Technical Editor

New data on thermophysical properties are being constantly accumulated at TPRC. Contact TPRC and use its interim updating services for the most current information.

MASTER INDEX
To Materials and Properties

Y. S. Touloukian
Director
Center for Information and Numerical Data Analysis and Synthesis
and
Distinguished Atkins Professor of Engineering
School of Mechanical Engineering
Purdue University
and
Visiting Professor of Mechanical Engineering
Auburn University

C. Y. Ho
Assistant Director
and
Head of Reference Data Division
Center for Information and Numerical Data Analysis and Synthesis
Purdue University

IFI/PLENUM • NEW YORK-WASHINGTON

202045

Library of Congress Cataloging in Publication Data

Touloukian, Yeram Sarkis, 1918-
  Master index to materials and properties.

  (Thermophysical properties of matter; v. [14])
  1. Matter — Properties — Indexes. I. Ho, Cho Yen,
1928-   joint author. II. Title.
QC173.397.P87 1970 vol. 14 [QC1733]   536'.08s
ISBN 0-306-67092-5   [541'.042'1]   79-11021

ISBN (Set) 0-306-67020-8

IFI/Plenum Data Company is a division of
Plenum Publishing Corporation
227 West 17th Street, New York, N.Y. 10011

Distributed in Europe by Heyden & Son, Ltd.
Spectrum House, Alderton Crescent
London NW4 3XX, England

Printed in the United States of America

"In this work, when it shall be found that much is omitted, let it not be forgotten that much likewise is performed..."

Contents

Foreword to the Series

In 1957, the Thermophysical Properties Research Center (TPRC) of Purdue University, under the leadership of its founder, Professor Y. S. Touloukian, began to develop a coordinated experimental, theoretical, and literature review program covering a set of properties of great importance to science and technology. Over the years, this program has grown steadily, producing bibliographies, data compilations and recommendations, experimental measurements, and other output. The series of volumes for which these remarks constitute a foreword is one of these many important products. These volumes are a monumental accomplishment in themselves, requiring for their production the combined knowledge and skills of dozens of dedicated specialists. The Thermophysical Properties Research Center deserves the gratitude of every scientist and engineer who uses these compiled data.

The individual nontechnical citizen of the United States has a stake in this work also, for much of the science and technology that contributes to his well-being relies on the use of these data. Indeed, recognition of this importance is indicated by a mere reading of the list of the financial sponsors of the Thermophysical Properties Research Center; leaders of the technical industry of the United States and agencies of the Federal Government are well represented.

Experimental measurements made in a laboratory have many potential applications. They might be used, for example, to check a theory, or to help design a chemical manufacturing plant, or to compute the characteristics of a heat exchanger in a nuclear power plant. The progress of science and technology demands that results be published in the open literature so that others may use them. Fortunately for progress, the useful data in any single field are not scattered throughout the tens of thousands of technical journals published throughout the world. In most fields, fifty percent of the useful work appears in no more than thirty or forty journals. However, in the case of TPRC, its field is so broad that about 100 journals are required to yield fifty percent. But that other fifty percent! It is scattered through more than 3500 journals and other documents, often items not readily identifiable or obtainable. Over 85,000 references are now in the files.

Thus, the man who wants to use existing data, rather than make new measurements himself, faces a long and costly task if he wants to assure himself that he has found all the relevant results. More often than not, a search for data stops after one or two results are found—or after the searcher decides he has spent enough time looking. Now with the appearance of these volumes, the scientist or engineer who needs these kinds of data can consider himself very fortunate. He has a single source to turn to; thousands of hours of search time will be saved, innumerable repetitions of measurements will be avoided, and several billions of dollars of investment in research work will have been preserved.

However, the task is not ended with the generation of these volumes. A critical evaluation of much of the data is still needed. Why are discrepant results obtained by different experimentalists? What undetected sources of systematic error may affect some or even all measurements? What value can be derived as a "recommended" figure from the various conflicting values that may be reported? These questions are difficult to answer, requiring the most sophisticated judgment of a specialist in the field. While a number of the volumes in this Series do contain critically evaluated and recommended data, these are still in the minority. The data are now being more intensively evaluated by the staff of TPRC as an integral part of the effort of the National Standard Reference Data System (NSRDS). The task of the National Standard Reference Data System is to organize and operate a comprehensive program to prepare compilations of critically evaluated data on the properties of substances. The NSRDS is administered by the National Bureau of Standards under a directive from the Federal Council for Science and Technology, augmented by special legislation of the Congress of the United States. TPRC is one of the national resources participating in the National Standard Reference Data System in a united effort to satisfy the needs of the technical community for readily accessible, critically evaluated data.

As a representative of the NBS Office of Standard Reference Data, I want to congratulate Professor Touloukian and his colleagues on the accomplishments represented by this Series of reference data

books. Scientists and engineers the world over are indebted to them. The task ahead is still an awesome one and I urge the nation's private industries and all concerned Federal agencies to participate in fulfilling this national need of assuring the availability of standard numerical reference data for science and technology.

EDWARD L. BRADY
Associate Director for Information Programs
National Bureau of Standards

Preface to the Series

Thermophysical Properties of Matter, the TPRC Data Series, is the culmination of twenty years of pioneering effort of the Thermophysical Properties Research Center (TPRC), one of the operating centers of the Center for Information and Numerical Data Analysis and Synthesis (CINDAS) at Purdue University, in the generation of tables of numerical data for science and technology. It constitutes the restructuring, accompanied by extensive revision and expansion of coverage, of the original *TPRC Data Book*, first released in 1960 in loose-leaf format, 11″ × 17″ in size, and issued in June and December annually in the form of supplements. The original loose-leaf *Data Book* was organized in three volumes: (1) metallic elements and alloys; (2) nonmetallic elements, compounds, and mixtures which are solid at N.T.P., and (3) non-metallic elements, compounds, and mixtures which are liquid or gaseous at N.T.P. Within each volume, each property constituted a chapter.

Because of the vast proportions the *Data Book* began to assume over the years of its growth and the greatly increased effort necessary in its maintenance by the user, it was decided in 1967 to change from the loose-leaf format to a conventional publication. Thus, the December 1966 supplement of the original *Data Book* was the last supplement disseminated by TPRC.

While the manifold physical, logistic, and economic advantages of the bound volume over the loose-leaf oversize format are obvious and welcome to all who have used the unwieldy original volumes, the assumption that this work will no longer be kept on a current basis because of its bound format would not be correct. Fully recognizing the need of many important research and development programs which require the latest available information, TPRC has instituted a *Data Update Plan* enabling the subscriber to inquire, by telephone if necessary, for specific information and receive, in many instances, same-day response on any new data processed or revision of published data since the latest edition. In this context, the TPRC Data Series departs drastically from the conventional handbook and giant multivolume classical works, which are no longer adequate media for the dissemination of numerical data of science and technology without a continuing activity on contemporary coverage. The loose-leaf arrangements of many works fully recognize this fact and attempt to develop a combination of bound volumes and loose-leaf supplement arrangements as the work becomes increasingly large. TPRC's *Data Update Plan* is indeed unique in this sense since it maintains the contents of the TPRC Data Series current and live on a day-to-day basis between editions. In this spirit, I strongly urge all purchasers of these volumes to complete in detail and return the *Volume Registration Certificate* which accompanies each volume in order to assure themselves of the continuous receipt of annual listing of corrigenda during the life of the edition.

The TPRC Data Series consists of 13 independent volumes. The first seven of these volumes were published in 1970, Volumes 8 and 9 in 1972, Volume 10 in 1973, Volumes 11 and 12 in 1975, and a Supplement to Volume 6 and Volume 13 in 1976. The organization of the TPRC Data Series makes each volume a self-contained entity available individually without the need to purchase the entire Series.

The coverage of the specific thermophysical properties represented by this Series constitutes the most comprehensive and authoritative collection of numerical data of its kind for science and technology.

Whenever possible, a uniform format has been used in all volumes, except when variations in presentation were necessitated by the nature of the property or the physical state concerned. In spite of the wealth of data reported in these volumes, it should be recognized that all volumes are not of the same degree of completeness. However, as additional data are processed at TPRC on a continuing basis, subsequent editions will become increasingly more complete and up to date. Each volume in the Series basically comprises three sections, consisting of a text, the body of numerical data with source references, and a material index.

The aim of the textual material is to provide a complementary or supporting role to the body of numerical data rather than to present a treatise on the subject of the property. The user will find a basic theoretical treatment, a comprehensive presentation of selected works which constitute reviews, or compendia of empirical relations useful in estimation of the property when there exists a paucity of data or when data are completely lacking. Established major experimental techniques are also briefly reviewed.

The body of data is the core of each volume and is presented in both graphical and tabular formats for convenience of the user. Every single point of numerical data is fully referenced as to its original source and no secondary sources of information are used in data extraction. In general, it has not been possible to critically scrutinize all the original data presented in these volumes, except to eliminate perpetuation of gross errors. However, in a significant number of cases, such as for the properties of liquids and gases, the thermal conductivity and thermal diffusivity of all the elements, and the thermal expansion of most materials in all material categories, the task of full evaluation, synthesis, and correlation has been completed. It is hoped that in subsequent editions of this continuing work, not only new information will be reported but the critical evaluation will be extended to increasingly broader classes of materials and properties.

The third and final major section of each volume is the material index. This is the key to the volume, enabling the user to exercise full freedom of access to its contents by any choice of substance name or detailed alloy and mixture composition, trade name, synonym, etc. Of particular interest here is the fact that in the case of those properties which are reported in separate companion volumes, the material index in each of the volumes also reports the contents of the other companion volumes.* The sets of companion volumes are as follows:

Thermal conductivity:	Volumes 1, 2, 3
Specific heat:	Volumes 4, 5, 6, 6S
Radiative properties:	Volumes 7, 8, 9
Thermal expansion:	Volumes 12, 13

*For the first edition of the Series, this arrangement was not feasible for Volumes 6S, 7, 8, and 12 due to the sequence and schedule of their publication, and the Supplement to Volume 6 (Volume 6S) carries its own independent material index.

The ultimate aims and functions of TPRC's Data Tables Division are to extract, evaluate, reconcile, correlate, and synthesize all available data for the thermophysical properties of materials with the result of obtaining internally consistent sets of property values, termed the "recommended reference values." In such work, gaps in the data often occur, for ranges of temperature, composition, etc. Whenever feasible, various techniques are used to fill in such missing information, ranging from empirical procedures to detailed theoretical calculations. Such studies are resulting in valuable new estimation methods being developed which have made it possible to estimate values for substances and/or physical conditions presently unmeasured or not amenable to laboratory investigation. Depending on the available information for a particular property and substance, the end product may vary from simple tabulations of isolated values to detailed tabulations with generating equations, plots showing the concordance of the different values, and, in some cases, over a range of parameters presently unexplored in the laboratory.

The TPRC Data Series constitutes a permanent and valuable contribution to science and technology. These constantly growing volumes are invaluable sources of data to engineers and scientists, sources in which a wealth of information heretofore unknown or not readily available has been made accessible. We look forward to continued improvement of both format and contents so that TPRC may serve the scientific and technological community with ever-increasing excellence in the years to come. In this connection, the staff of TPRC is most anxious to receive comments, suggestions, and criticisms from all users of the volumes. An increasing number of colleagues are making available at the earliest possible moment reprints of their papers and reports as well as pertinent information on the more obscure publications. I wish to renew my earnest request that this procedure become a universal practice since it will prove to be most helpful in making TPRC's continuing effort more complete and up to date.

It is indeed a pleasure to acknowledge with gratitude the multisource financial assistance received from over fifty sponsors which has made the continued generation of these tables possible. In particular, I wish to single out the sustained major support received from the Air Force Materials Laboratory–Air Force Systems Command, the Defense Supply Agency, the Office of Standard Reference Data–National Bureau of Standards, and the Office of Advanced Research and Technology–National Aeronautics and Space Administration.

TPRC is indeed proud to have been designated as a National Information Analysis Center for the Department of Defense as well as a component of the National Standard Reference Data System under the cognizance of the National Bureau of Standards.

While the preparation and continued maintenance of this work is the responsibility of TPRC's Data Tables Division, it would not have been possible without the direct input of TPRC's Scientific Documentation Division and, to a lesser degree, the Theoretical and Experimental Research Divisions. The authors of the various volumes are the Senior staff members in responsible charge of the work. It should be clearly understood, however, that many have contributed over the years and their contributions are specifically acknowledged in each volume. I wish to take this opportunity to personally thank those members of the staff, assistant researchers, graduate research assistants, and supporting graphics and technical typing personnel without whose diligent and painstaking efforts this work could not have materialized.

Y. S. TOULOUKIAN

Director
Center for Information and Numerical
Data Analysis and Synthesis
Distinguished Atkins Professor of Engineering

Purdue University
West Lafayette, Indiana
November 1978

Introduction to Index Volume

Now that the 13-volume *Thermophysical Properties of Matter – The TPRC Data Series* is completed, it may be quite appropriate and informative to have an overview of this Series which was published over the years 1970 to 1977. Statistical data on the Series, presented in the accompanying table, are self-explanatory and give an indication of the scope of this monumental work.

This 178-page Master Index to the 6362 individual materials and properties reported in the 13-volume Data Series is prepared with a dual purpose in mind. First, it will assist those who wish to use this work to rapidly ascertain whether a particular property for a given substance is reported in the Series, and if so, on which page of which volume it is to be found. Secondly, and perhaps equally important, it will serve as a reference source for those who do not have this Series to determine if the property and substance or material of interest is covered by this encyclopedic work.

Naturally, each of the 14 books of the 13-volume Series (including the Supplement to Volume 6) has its own materials index. However, because of the unusually large size of the Series it was felt that an index to all

SUMMARY OF STATISTICAL DATA ON THERMOPHYSICAL PROPERTIES OF MATTER – THE TPRC DATA SERIES

	Number of pages	Number of data sets	Number of references	Number of materials
Volume 1: Thermal Conductivity – Metallic Elements and Alloys	1595	5539	1446	892
Volume 2: Thermal Conductivity – Nonmetallic Solids	1302	4627	1037	812
Volume 3: Thermal Conductivity – Nonmetallic Liquids and Gases	707	1505	1406	170
Volume 4: Specific Heat – Metallic Elements and Alloys	830	1186	789	322
Volume 5: Specific Heat – Nonmetallic Solids	1737	1009	518	550
Volume 6: Specific Heat – Nonmetallic Liquids and Gases	383	863	665	56
Volume 6 Supplement	169	726	878	307
Volume 7: Thermal Radiative Properties – Metallic Elements and Alloys	1644	5130	520	242
Volume 8: Thermal Radiative Properties – Nonmetallic Solids	1890	4971	576	782
Volume 9: Thermal Radiative Properties – Coatings	1569	5269	475	1161
Volume 10: Thermal Diffusivity	760	1733	568	445
Volume 11: Viscosity	801	1803	1595	188
Volume 12: Thermal Expansion – Metallic Elements and Alloys	1440	4253	872	672
Volume 13: Thermal Expansion – Nonmetallic Solids	1786	4990	1213	815
Totals	16,613	43,604	12,258	

14 volumes will add considerably to the ease of using these volumes as well as serve an integrating function.

In the preparation of the Master Index, the Editors had to reconcile what at first seemed to be certain inconsistencies among the index entries of individual volumes. In actuality, what seem to be editorial inconsistencies are the result of improvements in styling and in naming and grouping of materials that were introduced over the period of 1970 to 1977, when these fourteen volumes were published. Therefore, certain guidelines had to be adopted in the naming, styling, and alphabetization of this Master Index. These guidelines are summarized on the following pages.

It is hoped that no serious errors exist in this Master Index. Even though great care was exercised in its preparation, together with an attempt to insure consistency and appropriate cross indexing, it is possible that certain oversights may have occurred. Therefore, it is always advisable to look for materials under more than a single name entry whenever any ambiguity in naming exists.

As it is evident from the composition of the book, the index was prepared and formatted by a special text formatting computer program. This effort was the contribution of Dr. H. H. Li of the CINDAS staff, which the Editors wish to acknowledge.

Guidelines to Indexing and Alphabetization

1. General Rules for Alphabetization

a. Material names are arranged alphabetically and in increasing numerical order when relevant. However, hyphenated alphanumeric prefixes (e.g., *n-*, *p-*, *o-*, *iso-*, 2-, etc.) as well as Greek characters and notations such as +, −, or / used in names or in the styling of entries are ignored for alphabetization purposes. Whenever appropriate, prefixes to chemical name listings are given in italics.

b. Abbreviations or acronyms are written with no punctuation or spacing between letters and are considered as words for purposes of alphabetization (e.g., AISI, ASTM, SAE, etc.).

c. In alphabetizing a material with one or more modifier(s), the material name is separated from the modifier by a comma. Entries for the material without any modifier(s) are listed first, followed by entries for the same material name alphabetized according to the first modifier word. In the case of metals, when a national designation appears, this modifier is listed immediately after the material name.

d. No entries are listed under modifier or descriptor terms as lead words. The same holds for trade names in general except for those few which have crept into common usage (e.g., Teflon, Freon; also R numbers for refrigerants, etc.). Modifiers always follow the specific material names or the generic or material class names to which they are attached. Examples are:

> Aluminum oxide, Coors AD99
> Iron, cast
> Iron, gray
> Marble, black
> Marble, powder
> Steel, carbon
> Steel, stainless
> Paint, white
> etc.

e. In listing mixtures of solids or fluids, the constituent substances are ordered alphabetically. Cermets are an exception to this rule, the oxide or the compound always being listed first.

2. Listing of Inorganic Compounds

a. The convention of distinguishing a complex oxide from a salt is based on the criterion that the former designation is used when the electropositive element in the anion is a metal (e.g., calcium tungsten oxide, $CaO \cdot WO_3$, is the correct naming rather than calcium tungstate, $CaWO_4$).

b. In the case of inorganic compounds, some constituent elements have multiple valence states. As a consequence, for a given pair of elements, two or more compounds are formed according to their valences. For example, the following chemical combinations of Cr and Si occur:

Chromium monosilicide	$CrSi$
Chromium disilicide	$CrSi_2$
Trichromium silicide	Cr_3Si
Trichromium disilicide	Cr_3Si_2
Pentachromium trisilicide	Cr_5Si_3
Hexachromium silicide	Cr_6Si

In such a case the entry "chromium silicides" is used as a lead entry and the compounds within the group are listed by their chemical formula only, following the lead entry, i.e.,

> Chromium silicides:
> $CrSi$
> $CrSi_2$
> Cr_3Si
> Cr_3Si_2
> Cr_5Si_3
> Cr_6Si

3. Alloys and Steels

Alloys are listed in one or more of the following forms:

a. *By the name of the predominant alloying element* as the name of the alloy, followed by a listing of different combinations of constituents. Only the two major constituents are listed: i.e., $A + B + \Sigma Xi$. The notation ΣXi indicates the presence of additional lesser constituents. Binary alloys are listed first, followed by multiple alloys. Examples are:

Aluminum alloys:	Nickel alloys:
Al + Co	Ni + Al
Al + Cu + ΣXi	Ni + Cr
Al + Fe	Ni + Cr + ΣXi
Al + Fe + ΣXi	Ni + Fe + ΣXi

Titanium alloys:
Ti + Al
Ti + Al + ΣXi

Together with each of the above entries one or more alloy designations may appear with modifiers such as: country of origin, alloy number, trade name, work or heat treatment, etc.

b. *Under AISI, ASTM, and SAE designations,* e.g.,

AISI 310, stainless steel
ASTM B265-58T, titanium alloy

In the above two cases the entries are also listed as

Steel, stainless, AISI 310
and
Titanium alloy, ASTM B265-58T

c. Alloys are also cross-referenced under the entry "Steel" or "Steel, stainless" for each alloy separately listed under specific designations such as AISI, ASTM, SAE, or by common trade name.

From the above, it is evident that it is often advisable to look for alloys under entries for alloys of the predominant constituent as well as under AISI, ASTM, or SAE designations and under the word "Steel." Well recognized special alloys are also cross-listed individually under their trade names (e.g., Alumel, Chromel, etc.).

4. Designation of Mixtures (Solid and Fluid), Cermets, and Intermetallic Compounds

a. *Mixtures of solids* of A + B + C + . . . are separated by a plus (+) sign and are ordered alphabetically by the constituents' names. Binary mixtures are listed first, followed by mixtures of increasing numbers of constituents. The word "mixture" appears at the end of the entry, preceded by a comma.

b. *Cermets* are listed both under the name of the oxide or the elemental compound as well as under the general entry "Cermets." A plus (+) sign is used in the ordering of the constituents, with the oxide or compound always listed first. The word "cermet" appears at the end of the entry, preceded by a comma.

c. *Intermetallic compounds* are listed both under their conventional chemical name followed by the words "Intermetallic Compound" as well as under the general heading of "Intermetallic Compounds." The constituents are separated by a dash (−).

d. *Fluid mixtures* of A−B−C−. . . are ordered in alphabetical order of their constituents, which are separated by a dash (−). The word "mixture" appears at the end of the entry. Binary mixtures are listed first, followed by mixtures of increasing numbers of constituents.

5. Grouping of Common Materials: Bricks, Cements, Ceramics, Composites, Concretes, Enamels, Glasses, Graphites, Oxide Mixtures, Porcelains, Refractories, Rubbers, and Polymers

a. Mixtures of oxides which are recognized as ceramics, enamels, glasses, porcelains, or refractories are listed under one of these lead names, followed by appropriate qualifiers. However, in the absence of uniform practice in the naming of such oxide mixtures, it is advisable to look under more than one name as well as under the listing of a given oxide or oxide mixture.

b. Concretes, graphites, and composites are also general grouping designations used as lead words to bring like materials together. These words are followed by a simple description of their components or appropriate modifiers.

6. Binders, Coatings, and Paints

These three terms are difficult to distinguish in practice in the search for materials data. They are all coatings which can be classified as pigmented coatings, contact coatings, and conversion coatings. For purposes of this index all coatings are listed under the lead words binder, coating, or paint in one of the following styles. Examples are:

Binder, 3M Kel-F 800 with zinc oxide pigment
or
Binder, 3M Kel-F 800 pigmented with:
Aluminum oxide
Sodium sulfate + Titanium oxide
Magnesium oxide
Titanium oxide
Zinc oxide
etc.
Coating, Acrylic on Aluminum substrate
or
Coating, Acrylic on:
Ceramic substrate
Epoxy substrate
Glass substrate
Polyurethane substrate
Stainless steel substrate
etc.

or   Paint, white velvet 3M

Paints, Fuller:
     D-70-6342
     Flat black decoret
     Flat black silicone
     Harvard Gray No. 2946
       etc.

It is indeed hoped that this Master Index volume to the TPRC Data Series on thermophysical properties of materials will prove helpful in many ways to all seekers for numerical data by serving as a master key to a vast collection of mostly evaluated data.

Master Index to Materials and Properties

Substance Name	Thermal Conductivity		Specif. Heat		Thermal Radiative Properties								Thermal Diffusivity		Viscosity		Thermal Expansion	
					Emissivity		Reflectivity		Absorptivity		Transmissiv.							
	V.	Page	V.	Page	V.	Page	V.	Page	V.	Page	V.	Page	V.	Page	V.	Page	V.	Page
Acetaldehyde		–	6s	1		–		–		–		–		–		–		–
Acetamidophenol		–	6s	53		–		–		–		–		–		–		–
Acetaminophenol		–	6s	53		–		–		–		–		–		–		–
Acetic acid		–	6s	1		–		–		–		–		–		–		–
Acetic ester		–	6s	35		–		–		–		–		–		–		–
Acetic ether		–	6s	35		–		–		–		–		–		–		–
Acetone	3	129	6	113		–		–		–		–		–	11	98		–
Acetone-benzene, mixture	3	440		–		–		–		–		–		–		–		–
Acetylaminophenol		–	6s	53		–		–		–		–		–		–		–
Acetylene	3	133	6	117		–		–		–		–		–	11	100		–
Acetylene-air, mixture	3	381		–		–		–		–		–		–		–		–
Acetylene dichloride		–	6s	28		–		–		–		–		–		–		–
Acetylene tetrabromide		–	6s	90		–		–		–		–		–		–		–
Acetylene tetrachloride		–	6s	90		–		–		–		–		–		–		–
Acetylenogen		–	5	405		–		–		–		–		–		–		–
Acrylic		–		–		–		–		–		–	10	594		–		–
Adiprene C + lithafax		–		–		–		–		–		–		–		–	13	1520
ADP, ammonium dihydrogen phosphate	2	679		–		–		–		–	8	604		–		–		–
Aggregate, Sil-O-Cel coarse grade, diatomite	2	1112		–		–		–		–		–		–		–		–
Air	3	512	6	293		–		–		–		–	10	518	11	608		–
Air-ammonia, mixture	3	442		–		–		–		–		–		–	11	624		–
Air-argon-carbon dioxide, mixture		–		–		–		–		–		–		–	11	602		–
Air-argon-carbon dioxide-helium, mixture		–		–		–		–		–		–		–	11	600		–
Air-argon-carbon dioxide-metane, mixture		–		–		–		–		–		–		–	11	603		–
Air-argon-helium-methane, mixture		–		–		–		–		–		–		–	11	601		–
Air-carbon dioxide, mixture		–		–		–		–		–		–		–	11	614		–
Air-carbon dioxide-helium, mixture		–		–		–		–		–		–		–	11	604		–
Air-carbon dioxide-helium-methane, mixture		–		–		–		–		–		–		–	11	605		–
Air-carbon dioxide-methane, mixture		–		–		–		–		–		–		–	11	616		–
Air-carbon monoxide, mixture	3	383		–		–		–		–		–		–		–		–
Air-helium, mixture	3	318		–		–		–		–		–		–		–		–
Air-helium-methane, mixture		–		–		–		–		–		–		–	11	606		–
Air-hydrogen chloride, mixture		–		–		–		–		–		–		–	11	626		–
Air-hydrogen sulphide, mixture		–		–		–		–		–		–		–	11	628		–
Air-methane, mixture	3	385		–		–		–		–		–		–	11	617		–
Air-steam, mixture	3	464		–		–		–		–		–		–		–		–
AISI 11H steel		–		–		–		–		–		–		–		–	12	1146
AISI 202 steel		–		–		–		–		–		–	10	339 340		–		–
AISI 301 stainless steel	1	1165	4	693	7	1221 1226	7	1269 1288	7	1300		–	10	345 348		–	12	1138 1141 1142
AISI 301 stainless steel, corrugated sheets		–		–		–		–		–		–	10	552		–		–

Substance Name	Thermal Conduc- tivity		Specif. Heat		Thermal Radiative Properties								Thermal Diffu- sivity		Visco- sity		Thermal Expan- sion	
					Emis- sivity		Reflec- tivity		Absorp- tivity		Trans- missiv.							
	V.	Page	V.	Page	V.	Page	V.	Page	V.	Page	V.	Page	V.	Page	V.	Page	V.	Page
AISI 302 stainless steel	1	1161	–		7	1212 1213	–		7	1291	–		10	345	–		12	1138 1142
AISI 303 stainless steel	1	1165 1168	–		7	1212 1226 1254 1258 1259 1260	–		7	1297	–		–		–		12	1138 1142
AISI 304 stainless steel	1	1161 1165 1168	4	699	7	1213 1227 1244	7	1270	–		–		–		–		–	
AISI 304ELC stainless steel	–		–		7	1213	–		–		–		–		–		–	
AISI 304L stainless steel	–		–		–		–		–		–		–		–		12	1138 1142
AISI 305 stainless steel	–		4	702	–		–		–		–		–		–		–	
AISI 309 stainless steel	–		–		–		–		–		–		10	346	–		–	
AISI 310 stainless steel	1	1167 1168	4	705	7	1212 1213	–		–		–		–		–		12	1138 1142
AISI 316 stainless steel	1	1165 1166	4	708	7	1221 1224 1237 1244	7	1266 1270 1271 1288	7	1300 130	–		10	347 348	–		12	1138 1143
AISI 321 stainless steel	–		–		7	1224 1237 1238 1244 1246	7	1266 1270 1272 1285	7	1294 1302	–		10	347	–		12	1138 1143
AISI 330 stainless steel	–		–		–		–		–		–		–		–		12	1177
AISI 347 stainless steel	1	1165 1166 1168	4	711	7	1212 1222	7	1288	–		–		10	348	–		12	1138 1143 1144
AISI 403 stainless steel	1	1149	–		–		–		–		–		–		–		–	
AISI 406 stainless steel	–		–		–		–		–		–		–		–		12	1138 1144
AISI 410 stainless steel	1	1150	–		–		–		–		–		10	340	–		12	1138 1144
AISI 416 stainless Steel	–		–		–		–		–		–		10	340 341	–		12	1138 1144
AISI 420 stainless steel	1	1162	4	678	–		–		–		–		–		–		12	1138 1144
AISI 422 stainless steel	–		–		–		–		–		–		–		–		12	1138 1145
AISI 430 stainless steel	1	1154	4	681	7	1193	–		–		–		10	341	–		–	
AISI 430F stainless steel	–		–		–		–		–		–		–		–		12	1138 1145
AISI 440C stainless steel	1	1154	–		–		–		–		–		–		–		12	1138 1145
AISI 446 stainless steel	1	1155 1156	4	684	7	1180 1187	7	1198	7	1207	–		10	341 342	–		12	1138 1145
AISI 455 stainless steel	–		–		–		–		–		–		–		–		12	1145
AISI 633 stainless steel	–		–		–		–		–		–		–		–		12	1145
AISI 1010 steel	1	1185	–		–		–		–		–		–		–		–	
AISI 1010C steel	1	1183	–		–		–		–		–		–		–		12	1167
AISI 1015C steel	1	1186	–		–		–		–		–		–		–		–	
AISI 1018 steel	–		–		–		–		–		–		10	358	–		–	
AISI 1020C steel	1	1183	–		–		–		–		–		–		–		–	

Substance Name	Thermal Conductivity		Specif. Heat		Thermal Radiative Properties										Thermal Diffusivity		Viscosity		Thermal Expansion	
					Emissivity		Reflectivity		Absorptivity		Transmissiv.									
	V.	Page	V.	Page	V.	Page	V.	Page	V.	Page	V.	Page	V.	Page	V.	Page	V.	Page
AISI 1045 steel		–		–		–		–		–		–	10	358		–		–
AISI 1095 steel	1	1114		–		–		–		–		–		–		–		–
AISI 2515 steel	1	1198 1199 1200		–		–		–		–		–		–		–		–
AISI 3140 steel		–		–		–		–		–		–	10	361		–		–
AISI 4130 steel	1	1153		–		–		–		–		–		–		–		–
AISI 4140 steel	1	1155		–		–		–		–		–		–		–		–
AISI 4340 steel	1	1213 1214		–		–		–		–		–		–		–	12	1177
Aldehyde		–	6s	1		–		–		–		–		–		–		–
Allene		–	6s	75		–		–		–		–		–		–		–
Allyl alcohol		–	6s	1		–		–		–		–		–		–		–
Allyl tribromide		–	6s	91		–		–		–		–		–		–		–
Allyl trichloride		–	6s	91		–		–		–		–		–		–		–
Alodine 401-45		–		–	9	1203	9	1208		–		–		–		–		–
Alodine 1200		–		–	9	1234		–		–		–		–		–		–
Alum	2	688		–		–		–		–		–		–		–		–
Alumina	2	99 106	5	26		–		–		–		–		–		–		–
Alumina, alpha		–		–		–		–		–		–		–		–	13	179
Alumina, hi	2	99		–		–		–		–		–		–		–		–
Alumina, ignited	2	106		–		–		–		–		–		–		–		–
Alumina, morganite		–		–		–		–		–		–		–		–	13	180
Alumina + mullite, mixture	2	322 335		–		–		–		–		–		–		–		–
Aluminum, Al	1	1	4	1	7	2 8 12 15	7	18 20 24 34 38 40	7	42 45 47 50 52 55	7	57	10	2		–	12	2
Aluminum, 1 percent impurities	1	925		–		–		–		–		–		–		–		–
Aluminum, anodized		–		–	9	1195 1199 1203	9	1207 1216 1219 1220 1222	9	1224 12		–		–		–		–
Aluminum, anodized Al 1199		–		–	9	1195	9	1207 1219 1220 1222	9	1226 1230 1231		–		–		–		–
Aluminum, Cockron home foil		–		–	7	4 5		–	7	42 43		–		–		–		–
Aluminum, foil, Kaiser		–		–	7	4 5		–	7	42		–		–		–		–
Aluminum, Hurwich home foil		–		–	7	4 5		–	7	42 43		–		–		–		–
Aluminum, SAP, sintered powder		–		–		–		–		–		–		–		–	12	4
Aluminum, soft		–		–		–		–		–		–		–		–	12	4
Aluminum, superaffinal		–		–		–		–		–		–		–		–	12	7

Substance Name	Thermal Conductivity		Specif. Heat		Thermal Radiative Properties								Thermal Diffusivity		Viscosity		Thermal Expansion	
					Emissivity		Reflectivity		Absorptivity		Transmissiv.							
	V.	Page	V.	Page	V.	Page	V.	Page	V.	Page	V.	Page	V.	Page	V.	Page	V.	Page
Aluminum alloys:																		
Al + Ag	–		–		–		7	896	–		–		–		–		–	
Al + Be	–		–		–		–		–		–		–		–		12	630
Al + Co	–		–		–		7	887	–		–		–		–		–	
Al + Cu	1	470	–		–		–		–		–		–		–		12	634
Al + Cu, Al 2014, anodized	–		–		9	1234	–		–		–		–		–		–	
Al + Cu, XB-18S	–		–		–		–		–		–		–		–		12	1016
Al + Fe	1	474	–		–		–		–		–		–		–		12	641
Al + Fe, DIN 712	1	475	–		–		–		–		–		–		–		–	
Al + Mg	1	477	–		–		7	890	–		–		–		–		12	646
Al + Mg, A54S	1	478 909	–		–		–		–		–		–		–		–	
Al + Mg, anodized	–		–		–		9	1254 1255	–		–		–		–		–	
Al + Mg, Magnalium	1	478	–		–		–		–		–		–		–		–	
Al + Mn	1	911	–		7	1112 1114	–		–		–		10	224	–		12	653
Al + Mo	–		–		–		–		–		–		–		–		12	654
Al + Ni	–		–		–		–		–		–		–		–		12	657
Al + Sb	1	469	–		–		–		–		–		–		–		–	
Al + Si	1	480	–		–		7	893	–		–		–		–		12	658
Al + Si, Al 2358	1	481	–		–		–		–		–		–		–		–	
Al + Si, Alpax	1	481	–		–		–		–		–		–		–		–	
Al + Si, Alusil	1	481	–		–		–		–		–		–		–		–	
Al + Sn	1	483	–		–		–		–		–		–		–		–	
Al + U	1	484	–		–		–		–		–		–		–		12	664
Al + Zn	1	487	–		–		–		–		–		–		–		12	666
Al + Zr	–		–		–		–		–		–		–		–		12	669
Al + Be + ΣXi	–		–		–		–		–		–		–		–		12	1006
Al + Cu + ΣXi	1	895	4	511	7	1062 1066 1072	7	1076 1083	7	1086	–		10	270	–		12	1011
Al + Cu + ΣXi, Al 2014	1	901	–		–		–		–		–		–		–		12	1011 1014
Al + Cu + ΣXi, Al 2017	–		–		–		–		–		–		–		–		12	1011 1014
Al + Cu + ΣXi, Al 2018	–		–		–		–		–		–		–		–		12	1011 1013
Al + Cu + ΣXi, Al 2020	–		–		–		–		–		–		–		–		12	1011 1014
Al + Cu + ΣXi, Al 2024	1	896 898 901	4	511	7	1063 1074	–		7	1087	–		–		–		12	1011 1014 1015
Al + Cu + ΣXi, Al 2024, anodized	–		–		9	1235	9	1237 1240	9	1241	–		–		–		–	
Al + Cu + ΣXi, Al 2024-T	–		–		7	1063 1068 1069	7	1078 1084	7	1087	–		–		–		–	
Al + Cu + ΣXi, Al 2024-T3	–		–		–		–		–		–		–		–		12	1015

Substance Name	Thermal Conductivity		Specif. Heat		Emissivity		Reflectivity		Absorptivity		Transmissiv.		Thermal Diffusivity		Viscosity		Thermal Expansion	
	V.	Page	V.	Page	V.	Page	V.	Page	V.	Page	V.	Page	V.	Page	V.	Page	V.	Page
Aluminum alloys: (continued)																		
Al + Cu + ΣXi, Al 2024-T4		–		–		–		–		–		–		–		–	13	1013 1014
Al + Cu + ΣXi, Al 2025		–		–		–		–		–		–		–		–	12	1011 1013
Al + Cu + ΣXi, Al 2117		–		–		–		–		–		–		–		–	12	1015
Al + Cu + ΣXi, Al 2219		–		–		–	7	1078		–		–		–		–	12	1015
Al + Cu + ΣXi, Al 2618		–		–		–		–		–		–		–		–	12	1015
Al + Cu + ΣXi, Alclad 24S C		–		–		–	7	1084		–		–		–		–		–
Al + Cu + ΣXi, Alclad 24S T		–		–	7	1068 1069		–		–		–		–		–		–
Al + Cu + ΣXi, Alclad 20240		–		–		–	7	1084		–		–		–		–		–
Al + Cu + ΣXi, Alclad 2024T		–		–	7	1068 1069		–		–		–		–		–		–
Al + Cu + ΣXi, Alcoa 12S	1	897 899 900		–		–		–		–		–		–		–		–
Al + Cu + ΣXi, Alcoa 14S	1	901		–		–		–		–		–		–		–	12	1014
Al + Cu + ΣXi, Alcoa 17S		–		–		–		–		–		–		–		–	12	1014
Al + Cu + ΣXi, Alcoa 18S		–		–		–		–		–		–		–		–	12	1013
Al + Cu + ΣXi, Alcoa 24S	1	896 898 901	4	512		–		–		–		–		–		–	12	1014 1015
Al + Cu + ΣXi, Alcoa 24S T		–		–	7	1063 1068 1069	7	1078 1084	7	1087		–		–		–		–
Al + Cu + ΣXi, Alcoa 25S T4		–		–		–		–		–		–		–		–	12	1013 1014
Al + Cu + ΣXi, Alcoa 25S		–		–		–		–		–		–		–		–	12	1013
Al + Cu + ΣXi, Alcoa A17S		–		–		–		–		–		–		–		–	12	1015
Al + Cu + ΣXi, anodized		–		–	9	1234 1235	9	1237 1240	9	1241		–		–		–		–
Al + Cu + ΣXi, British, 2L-11	1	900		–		–		–		–		–		–		–		–
Al + Cu + ΣXi, British, L-8	1	899		–		–		–		–		–		–		–		–
Al + Cu + ΣXi, British, Y-1	1	900		–		–		–		–		–		–		–		–
Al + Cu + ΣXi, British, Y-2	1	900		–		–		–		–		–		–		–		–
Al + Cu + ΣXi, duralumin	1	896		–		–		–		–		–		–		–	12	1015 1016
Al + Cu + ΣXi, German Y alloy	1	896 898		–		–		–		–		–		–		–		–
Al + Cu + ΣXi, Japanese, 2E-8	1	899		–		–		–		–		–		–		–		–
Al + Cu + ΣXi, Japanese M-1	1	899		–		–		–		–		–		–		–		–
Al + Cu + ΣXi, K. S. special alloy	1	902		–		–		–		–		–		–		–		–
Al + Cu + ΣXi, L72		–		–		–	7	1084	7	1087		–		–		–		–
Al + Cu + ΣXi, Nelson-kBbenleg 10	1	896		–		–		–		–		–		–		–		–
Al + Cu + ΣXi, RR 53	1	901		–		–		–		–		–		–		–		–
Al + Cu + ΣXi, RR 59	1	898		–		–		–		–		–		–		–		–
Al + Cu + ΣXi, Y-alloy	1	896 898		–		–		–		–		–		–		–	12	1011 1016

8

Substance Name	Thermal Conductivity V.	Page	Specif. Heat V.	Page	Emissivity V.	Page	Reflectivity V.	Page	Absorptivity V.	Page	Transmissiv. V.	Page	Thermal Diffusivity V.	Page	Viscosity V.	Page	Thermal Expansion V.	Page
Aluminum alloys: (continued)																		
Al + Fe + ΣXi	1	905	–		–		7	1090 1094	–		–		10	273	–		12	1027
Al + Fe + ΣXi, Al 1060	–		–		–		–		–		–		–		–		12	6
Al + Fe + ΣXi, Al 1075	–		–		–		7	28	–		–		–		–		–	
Al + Fe + ΣXi, Al 1100	1	906 920	–		7	10 16	–		–		–		–		–		12	641
Al + Fe + ΣXi, Alclad 2024, anodized	–		–		–		9	1244 1252	–		–		–		–		–	
Al + Fe + ΣXi, Alcoa 2S	1	906 920	–		–		–		–		–		–		–		12	641
Al + Fe + ΣXi, anodized	–		–		–		9	1243 1251 1252	9	1253	–		–		–		–	
Al + Fe + ΣXi, cond-Al	1	906	–		–		–		–		–		–		–		–	
Al + Fe + ΣXi, L-34	–		–		–		7	1092 1095	–		–		–		–		–	
Al + Fe + ΣXi, L-34, anodized	–		–		–		9	1252	9	1253	–		–		–		–	
Al + Fe + ΣXi, J51	1	906	–		–		–		–		–		–		–		–	
Al + Mg + ΣXi	1	908	–		7	1098 1100	7	1105	7	1110	–		10	276	–		12	1028
Al + Mg + ΣXi, Al 5052	1	478 909	–		–		–		–		–		–		–		12	1028 1030
Al + Mg + ΣXi, Al 5053-3	–		–		7	1101	–		–		–		–		–		–	
Al + Mg + ΣXi, Al 5083	1	909	–		–		–		–		–		–		–		12	1028 1030
Al + Mg + ΣXi, Al 5086	1	909	–		–		–		–		–		–		–		–	
Al + Mg + ΣXi, Al 5154	1	478 909	–		–		–		–		–		–		–		–	
Al + Mg + ΣXi, Al 5456	1	909	–		–		–		–		–		–		–		12	1028 1030
Al + Mg + ΣXi, Al 6053	–		–		–		–		–		–		–		–		12	1028 1030
Al + Mg + ΣXi, Al 6061	–		–		7	1098	7	1106	7	1110	–		–		–		12	1028 1030
Al + Mg + ΣXi, Al 6063	1	909	–		–		–		–		–		–		–		–	
Al + Mg + ΣXi, Alcoa 52S	1	478 909	–		–		–		–		–		–		–		–	
Al + Mg + ΣXi, Alcoa 53S O	–		–		7	1101	–		–		–		–		–		–	
Al + Mg + ΣXi, Alcoa 61S	–		–		–		–		–		–		–		–		12	1030
Al + Mg + ΣXi, Alcoa 63S	1	909	–		–		–		–		–		–		–		–	
Al + Mg + ΣXi, anodized	–		–		–		9	1254 1255	–		–		–		–		–	
Al + Mg + ΣXi, K186	1	909	–		–		–		–		–		–		–		–	
Al + Mg + ΣXi, LK183	1	909	–		–		–		–		–		–		–		–	
Al + Mg + ΣXi, RR 131D	1	909	–		–		–		–		–		–		–		–	
Al + Mn + ΣXi	–		–		–		–		–		–		–		–		12	1034
Al + Mn + ΣXi, Al 3003	1	912	–		–		–		–		–		–		–		12	1034 1036 ; 13	616

Substance Name	Thermal Conductivity		Specif. Heat		Thermal Radiative Properties										Thermal Diffusivity		Viscosity		Thermal Expansion	
					Emissivity		Reflectivity		Absorptivity		Transmissiv.									
	V.	Page	V.	Page	V.	Page	V.	Page	V.	Page	V.	Page	V.	Page	V.	Page	V.	Page	V.	Page
Aluminum alloys: (continued)																				
Al + Mn + ΣXi, Al 3004	1	912	–		–		–		–		–		–		–		–			
Al + Mn + ΣXi, Alcoa 3S	1	912	–		–		–		–		–		–		–		12	1034 1036		
Al + Mn + ΣXi, Alcoa 4S	1	912	–		–		–		–		–		–		–		–			
Al + Ni + ΣXi	1	914	–		–		–		–		–		–		–		12	1038		
Al + Ni + ΣXi, RAE 40C	1	915	–		–		–		–		–		–		–		12	1040		
Al + Ni + ΣXi, RAE 47B	1	915	–		–		–		–		–		–		–		12	1040		
Al + Ni + ΣXi, RAE 47D	1	915	–		–		–		–		–		–		–		12	1040		
Al + Ni + ΣXi, RAE 55	1	915	–		–		–		–		–		–		–		12	1040		
Al + Si + ΣXi	1	917	–		–		–		–		–		10	277	–		12	1042		
Al + Si + ΣXi, Al 132	1	919	–		–		–		–		–		–		–		–			
Al + Si + ΣXi, Al 356	–		–		–		–		–		–		–		–		12	1044		
Al + Si + ΣXi, Al 4032	–		–		–		–		–		–		–		–		12	1042 1044		
Al + Si + ΣXi, Al 6151	–		–		–		–		–		–		–		–		12	1042 1044		
Al + Si + ΣXi, Alpax gamma	1	918	–		–		–		–		–		–		–		12	1046		
Al + Si + ΣXi, K. S. alloy 245	1	920	–		–		–		–		–		–		–		–			
Al + Si + ΣXi, K. S. alloy 280	1	920	–		–		–		–		–		–		–		–			
Al + Si + ΣXi, Lo-ex	1	919	–		–		–		–		–		–		–		12	1046		
Al + Si + ΣXi, RR 50	1	918 919 920	–		–		–		–		–		–		–		12	1046		
Al + Si + ΣXi, RR 53C	1	918	–		–		–		–		–		–		–		12	1046		
Al + Si + ΣXi, SA1	1	918 919	..		–		–		–		–		–		–		12	1046		
Al + Si + ΣXi, SA14	–		–		–		–		–		–		–		–		12	1046		
Al + Si + ΣXi, SA44	1	918 919	–		–		–		–		–		–		–		–			
Al + Si + ΣXi, Tens-50	–		–		–		–		–		–		–		–		12	1044		
Al + Si + ΣXi, γ-Silumin modified	1	920	–		–		–		–		–		–		–		–			
Al + Zn + ΣXi	1	922	4	514	7	1116 1118 1121 1124	7	1128	7	1131	–		10	234	–		12	1051		
Al + Zn + ΣXi, Al 7039	–		–		–		–		–		–		–		–		12	1051 1053		
Al + Zn + ΣXi, Al 7075	1	923	4	514	7	1122	–		–		–		–		–		12	1051 1053		
Al + Zn + ΣXi, Al 7075, anodized	–		–		9	1256	9	1258	–		–		–		–		–			
Al + Zn + ΣXi, Al 7075-T	–		–		7	1119	7	1129	7	1132	–		–		–		–			
Al + Zn + ΣXi, Al 7075-T6	–		–		7	1126	–		–		–		–		–		12	1053		
Al + Zn + ΣXi, Al 7075-T73	–		–		–		–		–		–		–		–		12	1053		
Al + Zn + ΣXi, Al 7079	–		–		–		–		–		–		–		–		12	1051		
Al + Zn + ΣXi, Al 7079-T6	–		–		–		–		–		–		–		–		12	1053		
Al + Zn + ΣXi, Alclad 75S T	–		–		7	1116 1119	7	1129	–		–		–		–		–			

Substance Name	Thermal Conductivity V.	Page	Specif. Heat V.	Page	Emissivity V.	Page	Reflectivity V.	Page	Absorptivity V.	Page	Transmissiv. V.	Page	Thermal Diffusivity V.	Page	Viscosity V.	Page	Thermal Expansion V.	Page
Aluminum alloys: (continued)																		
Al + Zn + ΣXi, Alclad 7075T	-		-		7	1116 1119	7	1129	-		-		-		-		-	
Al + Zn + ΣXi, Alcoa 75S	1	923	4	514	-		-		-		-		-		-		12	1051 1053
Al + Zn + ΣXi, Alcoa 75S T	-		-		7	1119	7	1129	7	1132	-		-		-		-	
Al + Zn + ΣXi, Alcoa 75S T6	-		-		-		-		-		-		-		-		12	1053
Al + Zn + ΣXi, anodized	-		-		9	1256	9	1258	-		-		-		-		-	
Al + Zn + ΣXi, British, L-5	1	923	-		-		-		-		-		-		-		-	
Al + Zr + ΣXi, L5	-		-		-		-		-		-		-		-		12	1053
Al + Zn + ΣXi, RR 77	1	923	-		-		-		-		-		-		-		12	1053
Al + Fe + Si + ΣXi, Al 1075, anodized	-		-		-		9	1243	-		-		-		-		-	
Al + Fe + Si + ΣXi, Al 1145, anodized	-		-		-		9	1251	9	1253	-		-		-		-	
Aluminum antimonide, AlSb	-		5	297	-		8	1352	-		-		-		-		-	
Aluminum-antimony intermetallic compound, AlSb	-		-		-		-		-		-		-		-		12	414
Aluminum boride, AlB$_{12}$	-		-		8	732	-		-		-		-		-		-	
Aluminum borosilicate complex, natural	2	855	-		-		-		-		-		-		-		-	
Aluminum carbide + ΣXi, mixture	-		5	395	-		-		-		-		-		-		-	
Aluminum-copper intermetallic compounds:																		
AlCu	-		-		-		-		-		-		-		-		12	417 419
AlCu$_2$	-		-		-		-		-		-		-		-		12	417 420
AlCu$_3$	-		-		-		-		-		-		-		-		12	417 422
Al$_2$Cu	-		-		-		-		-		-		-		-		12	417 418
Al$_4$Cu$_9$	-		-		-		-		-		-		-		-		12	417 421
Aluminum fluosilicate, 2AlFO·SiO$_2$, Brazil topaz	2	251	-		-		-		-		-		-		-		-	
Aluminum foil	-		-		7	5 6 9	7	26 27 40	7	43 50 55	7	60	-		-		-	
Aluminum foil, Reynolds wrap	-		-		-		7	40	7	55	-		-		-		-	
Aluminum-gold intermetallic compound, Al$_2$Au	-		-		-		-		-		-		-		-		12	423
Aluminum-iron intermetallic compounds:																		
AlFe	-		-		-		-		-		-		-		-		12	433 435 436
AlFe$_3$	-		-		-		-		-		-		-		-		12	434 435 437
Al$_2$Fe	-		-		-		-		-		-		-		-		12	428 429 432
Al$_3$Fe	-		-		-		-		-		-		-		-		12	427 429 431

Substance Name	Thermal Conductivity		Specif. Heat		Thermal Radiative Properties										Thermal Diffusivity		Visco-sity		Thermal Expansion	
					Emissivity		Reflectivity		Absorptivity		Transmissiv.									
	V.	Page	V.	Page	V.	Page	V.	Page	V.	Page	V.	Page	V.	Page	V.	Page	V.	Page	V.	Page
Aluminum-iron intermetallic compounds: (continued)																				
Al₈Fe₂	-		-		-		-		-		-		-			-	12	426 429 430		
Aluminum-nickel intermetallic compounds:																				
AlNi	-		-		-		-		-		-		-			-	12	438 440		
AlNi₃	-		-		-		-		-		-		-			-	12	441 443		
Aluminum nitride, AlN	2	653	5	1075	8	1030 1031 1033	8	1035	-		-		-			-	13	1127		
Aluminum oxides:																				
Al₂O₃	-		5	26	8	141 142 146 148	8	157 163	8	166 168	8	169	10	378		-	13	176		
Al₂O₃, powder	2	1040	-		-		-		-		-		-			-	-			
AP-30	2	99	-		-		-		-		-		-			-	-			
AV-30	2	102	-		-		-		-		-		-			-	-			
B45F	2	101	-		-		-		-		-		-			-	-			
Coors AD85	-		-		8	150	-		-		8	172	-			-	-			
Coors AD94	-		-		8	150	-		-		8	172	-			-	-			
Coors AD96	-		-		8	150	-		-		8	172	-			-	-			
Coors AD99	-		-		8	144 147 150	-		-		8	172	-			-	-			
Coors AD995	-		-		8	150	-		-		-		-			-	-			
Coors BD96	-		-		8	206	-		-		8	214	-			-	-			
Coors BD98	-		-		8	206	-		-		8	214	-			-	-			
E98	2	101	-		-		-		-		-		-			-	-			
Gulton HS.B	2	103	-		-		-		-		-		-			-	-			
McDanel AP 35	-		-		8	151	-		-		8	172	-			-	-			
McDanel AV 30	-		-		8	151	-		-		8	172	-			-	-			
Norton 38-900	2	103 104	-		-		-		-		-		-			-	-			
Norton 5190	-		-		8	150	-		-		-		-			-	-			
Norton A 402	-		-		8	150	-		-		-		-			-	-			
Norton LA 603	-		-		8	144 147	8	160	8	168	-		-			-	-			
Norton RA 4213	-		-		8	144 147	8	160	8	168	-		-			-	-			
Ruby	-		-		-		8	174	-		8	176	-			-	-			
Sapphire	-		-		8	179 181 183	8	187	-		8	190	-			-	-			
TC 352, cintered	2	107	-		-		-		-		-		-			-	-			
Wesgo Al-300	2	101 107 108	-		-		-		-		-		-			-	-			

Substance Name	Thermal Conductivity		Specif. Heat		Thermal Radiative Properties								Thermal Diffusivity		Viscosity		Thermal Expansion	
					Emissivity		Reflectivity		Absorptivity		Transmissiv.							
	V.	Page	V.	Page	V.	Page	V.	Page	V.	Page	V.	Page	V.	Page	V.	Page	V.	Page
Aluminum oxide + aluminum, cermet	–		–		–		–		–		–		–		–		13	1306
Aluminum oxide + aluminum silicate, mixture	2	321	–		–		–		–		–		–		–		–	
Aluminum oxide + chromium, cermet	2	707	–		–		–		–		–		–		–		–	
Aluminum oxide + chromium + ΣXi, cermet	–		–		8	1355	–		–		–		–		–		13	1309
Aluminum oxide + chromium oxide, mixture	2	324	–		–		–		–		–		–		–		–	
Aluminum oxide + chromium oxide powder	–		–		8	554	–		–		–		–		–		–	
Aluminum oxide + magnesium oxide, mixture	–		–		–		–		–		–		10	429	–		–	
Aluminum oxide + manganese oxide, mixture	2	327 397	–		–		–		–		–		–		–		–	
Aluminum oxide + molybdenum, cermet	–		–		–		–		–		–		10	566	–		–	
Aluminum oxide + nickel aluminum alloy, cermet	–		–		8	1358 1359	8	1363	–		–		–		–		–	
Aluminum oxide + nickel oxide powder	–		–		8	556	–		–		–		–		–		–	
Aluminum oxide + silicon dioxide, mixture	2	328 402	–		–		–		–		–		10	426	–		–	
Aluminum oxide + silicon dioxide + ΣXi, mixture	2	453 487	5	1546	–		–		–		–		–		–		–	
Aluminum oxide + silicon oxide powder, mixture	–		–		8	558	8	560	–		–		10	431	–		–	
Aluminum oxide + titanium dioxide + ΣXi, mixture	2	456	–		–		–		–		–		–		–		–	
Aluminum oxide + tungsten + ΣXi, cermet	–		–		8	1375	–		–		–		–		–		–	
Aluminum oxide + zirconium dioxide, mixture	2	331 441	–		–		–		–		–		–		–		–	
Aluminum + oxygen, mixture	–		–		–		–		–		–		10	225	–		–	
Aluminum phosphate, $AlPO_4$	–		–		–		8	602	–		–		–		–		13	689
Aluminum phosphide, AlP	–		5	517	–		–		–		–		–		–		–	
Aluminum silicate + aluminum oxide, mixture	2	1090	–		–		–		–		–		–		–		–	
Aluminum silicates:	–		–		–		8	618	–		–		–		–		–	
Al_2SiO_5	–		5	1289	–		–		–		–		–		–		13	703
$Al_2Si_2O_7 \cdot 2H_2O$	–		5	1295	–		–		–		–		–		–		–	
Al_4SiO_8	–		–		–		–		–		–		–		–		13	703
$Al_6Si_2O_{13}$	2	254	5	1292	–		–		–		–		10	412	–		13	703
Aluminum–silver intermetallic compound, $AlAg_2$	–		–		–		–		–		–		–		–		12	444
Aluminum sulfates:																		
$Al_2(SO_4)_3$	–		5	1161	–		–		–		–		–		–		–	
$Al_2(SO_4)_3 \cdot 6H_2O$	–		5	1164	–		–		–		–		–		–		–	
Aluminum + tantalum aluminide powder	–		–		–		8	1431	–		–		–		–		–	
Aluminum titanium oxide, $Al_2O_3 \cdot TiO_2$	–		5	1298	–		–		–		–		–		–		12	548
Aluminum trifluoride, AlF_3	–		5	915	–		–		–		–		–		–		–	
Aluminum tungsten oxide, $2Al_2O_3 \cdot 5WO_3$	–		–		–		–		–		–		–		–		13	576

Substance Name	Thermal Conductivity		Specif. Heat		Thermal Radiative Properties								Thermal Diffusivity		Viscosity		Thermal Expansion	
					Emissivity		Reflectivity		Absorptivity		Transmissiv.							
	V.	Page	V.	Page	V.	Page	V.	Page	V.	Page	V.	Page	V.	Page	V.	Page	V.	Page
Aluminum-uranium intermetallic compound, Al₃U	–		–		–		–		–		–		–		–		12	447
Alundum	2	456	–		–		–		–		–		–		–		–	
Amalgam	1	216	–		–		–		–		–		–		–		–	
Amber, glass	2	924	–		–		–		–		–		–		–		–	
Aminobenzene	–		6s	1	–		–		–		–		–		–		–	
Aminomethane	–		6s	59	–		–		–		–		–		–		–	
1-Amino-2-nitrobenzene	–		6s	69	–		–		–		–		–		–		–	
1-Amino-3-nitrobenzene	–		6s	69	–		–		–		–		–		–		–	
1-Amino-4-nitrobenzene	–		6s	69	–		–		–		–		–		–		–	
2-Aminopropane	–		6s	57	–		–		–		–		–		–		–	
Ammonia, NH₃	3	95	6	61	–		–		–		–		–		11	68	–	
Ammonia, trideuterated	–		6s	1	–		–		–		–		–		–		–	
Ammonia-argon, mixture	–		–		–		–		–		–		–		11	342	–	
Ammonia-carbon monoxide, mixture	3	444	–		–		–		–		–		–		–		–	
Ammonia chloride, NH₄Cl	–		–		–		–		–		–		–		–		13	968
Ammonia-ethylene, mixture	3	446	–		–		–		–		–		–		11	514	–	
Ammonia-hydrogen, mixture	3	448	–		–		–		–		–		–		11	516	–	
Ammonia-hydrogen-nitrogen, mixture	3	500	–		–		–		–		–		–		–		–	
Ammonia-methane, mixture	–		–		–		–		–		–		–		11	526	–	
Ammonia-methylamine, mixture	–		–		–		–		–		–		–		11	540	–	
Ammonia nitrate, NH₄NO₃	–		–		–		–		–		–		–		–		13	671
Ammonia-nitrogen, mixture	3	451	–		–		–		–		–		–		11	531	–	
Ammonia-nitrous oxide, mixture	–		–		–		–		–		–		–		11	534	–	
Ammonia-oxygen, mixture	–		–		–		–		–		–		–		11	538	–	
Ammonium aluminum sulfates:																		
NH₄Al(SO₄)₂	–		5	1170	–		–		–		–		–		–		–	
NH₄Al(SO₄)₂·12H₂O	–		5	1173	–		–		–		–		–		–		–	
Ammonium dihydrogen arsenate, NH₄H₂AsO₄	–		–		–		–		–		8	604	–		–		13	616
Ammonium dihydrogen phosphate, NH₄H₂PO₄	2	679	–		–		–		–		–		–		–		13	689
Ammonium hydrogen sulfate, NH₄HSO₄	2	687	–		–		–		–		–		–		–		–	
Ammonium perchlorate, NH₄ClO₄	2	757	–		–		–		–		–		–		–		–	
Ammonium sulfate, (NH₄)₂SO₄	–		5	1167	–		–		–		–		–		–		–	
Amyl alcohol	–		6s	72	–		–		–		–		–		–		–	
pri-Amyl alcohol	–		6s	72	–		–		–		–		–		–		–	
sec-Amyl alcohol	–		6s	72	–		–		–		–		–		–		–	
tert-Amyl alcohol	–		6s	61	–		–		–		–		–		–		–	
Amyl carbinol	–		6s	44	–		–		–		–		–		–		–	
Amyldimethylmethane	–		6s	63	–		–		–		–		–		–		–	
α-Amylene	–		6s	73	–		–		–		–		–		–		–	
β-Amylene	–		6s	73	–		–		–		–		–		–		–	
Amylene hydrate	–		6s	61	–		–		–		–		–		–		–	
Anatase	–		–		–		8	464	–		–		–		–		13	395

Substance Name	Thermal Conductivity		Specif. Heat		Thermal Radiative Properties								Thermal Diffusivity		Viscosity		Thermal Expansion	
					Emissivity		Reflectivity		Absorptivity		Transmissiv.							
	V.	Page	V.	Page	V.	Page	V.	Page	V.	Page	V.	Page	V.	Page	V.	Page	V.	Page
Anglesite	–		–		–		–		–		8	1702	–		–		–	
Aniline	–		6s	1	–		–		–		–		–		–		–	
Aniline-benzyl acetate, mixture	–		–		–		–		–		–		–		11	543	–	
Anorthite	–		–		–		–		–		–		–		–		13	707
Anthracene	2	985	–		–		–		–		–		–		–		–	
Anthracin	2	985	–		–		–		–		–		–		–		–	
Anthracite	–		–		–		–		–		–		10	22	–		–	
Antimonic acid anyhdride	–		5	33	–		–		–		–		–		–		–	
Antimony, Sb	1	10	4	6	–		7	63	–		7	66	10	7	–		12	13
Antimony alloys:																		
Liquid state, 46.8 percent Sb	–		–		–		–		–		–		10	284	–		–	
Sb + Al	1	488	–		–		–		–		–		–		–		–	
Sb + Bi	1	489	–		–		–		–		–		–		–		12	675
Sb + Cd	1	492	–		–		–		–		–		–		–		–	
Sb + Pb	1	496	–		–		–		–		–		–		–		–	
Sb + Sn	1	497	–		–		–		–		–		–		–		–	
Sb + Be + ΣXi	1	926	–		–		–		–		–		–		–		–	
Antimony-gallium intermetallic compound, SbGa	–		–		–		–		–		–		–		–		12	450
Antimony-indium intermetallic compound, SbIn	–		–		–		–		–		–		–		–		12	455
Antimony-lanthanum intermetallic compound, SbLa	–		–		–		–		–		–		–		–		12	460
Antimony oxides:																		
Sb_2O_3	–		–		–		8	198	8	200	–		–		–		–	
Sb_2O_4	–		5	30	–		–		–		–		–		–		–	
Sb_2O_5	–		5	33	–		–		–		–		–		–		–	
Antimony oxide powder, KP	–		–		–		8	199	8	200	–		–		–		–	
Antimony selenide + antimony telluride + bismuth telluride, mixture	1	1392	–		–		–		–		–		–		–		–	
Antimony selenide + silver selenide + lead selenide, mixture	1	1379	–		–		–		–		–		10	480	–		–	
Antimony sulfide, Sb_2S_3	–		5	635	–		–		–		–		–		–		–	
Antimony sulfide + sulfur, mixture	–		–		–		–		–		–		10	519	–		–	
Antimony sulfur iodide, SbSI	–		5	485	–		–		–		–		–		–		–	
Antimony telluride, Sb_2Te_3	1	1241	–		–		–		–		–		–		–		–	
Antimony telluride + bismuth telluride, mixture	1	1388	–		–		–		–		–		–		–		–	
Antimony telluride + indium telluride, mixture	1	1386	–		–		–		–		–		–		–		–	
Anydrite	–		–		–		–		–		–		8	630	–		–	
Apple sauce	–		–		–		–		–		–		10	622	–		–	
Arcton 33	–		6s	29	–		–		–		–		–		–		–	
Argentum	1	340	4	208	–		–		–		–		–		–		–	
Argon, Ar	3	1	6	1	–		–		–		–		–		11	2	13	2
Argon-benzene, mixture	3	295	–		–		–		–		–		–		–		–	

Substance Name	Thermal Conductivity V.	Page	Specif. Heat V.	Page	Thermal Radiative Properties Emissivity V.	Page	Reflectivity V.	Page	Absorptivity V.	Page	Transmissiv. V.	Page	Thermal Diffusivity V.	Page	Viscosity V.	Page	Thermal Expansion V.	Page
Argon-carbon dioxide, mixture	3	297	–		–		–		–		–		–		11	285	–	
Argon-carbon dioxide-helium, mixture	–		–		–		–		–		–		–		11	581	–	
Argon-carbon dioxide-helium-methane, mixture	–		–		–		–		–		–		–		11	594	–	
Argon-carbon dioxide-methane, mixture	–		–		–		–		–		–		–		11	583	–	
Argon-deuterium, mixture	3	299	–		–		–		–		–		–		–		–	
Argon-deuterium-hydrogen-nitrogen, mixture	3	507	–		–		–		–		–		–		–		–	
Argon-deuterium-hydrogen-xenon, mixture	3	510	–		–		–		–		–		–		–		–	
Argon-deuterium-krypton, mixture	3	488	–		–		–		–		–		–		–		–	
Argon-deuterium-krypton-xenon, mixture	3	506	–		–		–		–		–		–		–		–	
Argon-deuterium-neon, mixture	3	490	–		–		–		–		–		–		–		–	
Argon-dimethyl ether, mixture	3	454	–		–		–		–		–		–		–		–	
Argon-dimethyl ether-propane, mixture	3	499	–		–		–		–		–		–		–		–	
Argon-helium, mixture	3	251	–		–		–		–		–		–		11	237	–	
Argon-helium-krypton, mixture	3	481	–		–		–		–		–		–		–		–	
Argon-helium-methane, mixture	–		–		–		–		–		–		–		11	582	–	
Argon-helium-neon, mixture	–		–		–		–		–		=		–		11	580	–	
Argon-helium-nitrogen, mixture	3	486	–		–		–		–		–		–		–		–	
Argon-helium-xenon, mixture	3	479	–		–		–		–		–		–		–		–	
Argon-hydrogen, mixture	3	301	–		–		–		–		–		–		11	289	–	
Argon-hydrogen-krypton, mixture	3	496	–		–		–		–		–		–		–		–	
Argon-hydrogen-krypton-xenon, mixture	3	505	–		–		–		–		–		–		–		–	
Argon-hydrogen-neon-nitrogen, mixture	3	509	–		–		–		–		–		–		–		–	
Argon-hydrogen-nitrogen, mixture	3	493	–		–		–		–		–		–		–		–	
Argon-hydrogen-nitrogen-oxygen, mixture	3	508	–		–		–		–		–		–		–		–	
Argon-krypton, mixture	3	263	–		–		–		–		–		–		11	249	–	
Argon-krypton-neon, mixture	3	478	–		–		–		–		–		–		–		–	
Argon-krypton-neon-xenon, mixture	3	504	–		–		–		–		–		–		–		–	
Argon-krypton-xenon, mixture	3	483	–		–		–		–		–		–		–		–	
Argon-methane, mixture	3	304	–		–		–		–		–		–		–		–	
Argon-methane-oxygen, mixture	3	485	–		–		–		–		–		–		–		–	
Argon-methanol, mixture	3	458	–		–		–		–		–		–		–		–	
Argon-neon, mixture	3	258	–		–		–		–		–		–		11	251	–	
Argon-nitrogen, mixture	3	306	–		–		–		–		–		–		11	294	–	
Argon-oxygen, mixture	3	311	–		–		–		–		–		–		–		–	
Argon-propane, mixture	3	316	–		–		–		–		–		–		–		–	
Argon-sulfur dioxide, mixture	–		–		–		–		–		–		–		11	348	–	
Argon-xenon, mixture	3	267	–		–		–		–		–		–		11	258	–	
Armalon, nonmetallic laminate	2	1032	–		–		–		–		–		⁓		–		–	
Armco 21-6-9	–		–		–		–		–		–		–		–		12	1148
Armco iron	1	157 158 159 160 161 163	4	102	7	303 308	7	322	7	332	–	10	84 95		–		12	160 163 164

Substance Name	Thermal Conductivity		Specif. Heat		Thermal Radiative Properties								Thermal Diffusivity		Viscosity		Thermal Expansion	
					Emissivity		Reflectivity		Absorptivity		Transmissiv.							
	V.	Page	V.	Page	V.	Page	V.	Page	V.	Page	V.	Page	V.	Page	V.	Page	V.	Page
Armco iron, oxidized		–		–	9	1297	9	1299		–		–		–		–		–
Aromatic polyamide		–		–		–		–		–		–		–		–	13	1393
Arsenic, As	1	15	4	9		–	8	3		–		–	10	9		–	13	7
Arsenic hydride		–	6s	2		–		–		–		–		–		–		–
Arsenic iodide		–	5	488		–		–		–		–		–		–		–
Arsenic oxides:																		
As$_2$O$_3$		–	5	36		–		–		–		–		–		–		–
As$_2$O$_5$		–	5	39		–		–		–		–		–		–		–
Arsenic selenide, As$_2$Se$_3$		–		–		–	8	1130		–	8	1133		–		–	13	1192
Arsenic sulfides:																		
AsS		–	5	638		–		–		–		–		–		–		–
As$_2$S$_3$		–	5	641		–	8	1177		–	8	1179		–		–		–
Arsenic telluride, As$_2$Te$_3$	1	1244		–		–		–		–		–		–		–		–
Arsenic trideuteride, AsD$_3$		–	6s	2		–		–		–		–		–		–		–
Arsenic triiodide, AsI$_3$		–	5	488		–		–		–		–		–		–		–
Arsine		–	6s	2		–		–		–		–		–		–		–
Arsine, trideuterated		–	6s	2		–		–		–		–		–		–		–
Asbestos cement board	2	1107		–		–		–		–		–	10	568		–		–
Asbestos fiber	2	1135		–		–		–		–		–		–		–		–
Ash	2	1059		–		–		–		–		–		–		–		–
Ash, volcanic	2	856		–		–		–		–		–		–		–		–
Ashes + dolomite + quartz sand, mixture		–		–		–		–		–		–	10	433		–		–
ASTM B80 HZ-32A, magnesium alloy		–		–		–		–		–		–		–		–	12	1208 1212
ASTM B80 ZH-62A, magnesium alloy		–		–		–		–		–		–		–		–	12	1213
ASTM B90 HM-21A, magnesium alloy		–		–		–		–		–		–		–		–	12	1212
ASTM B265-58T, grade 3 and 4, titanium alloy		–	4	257		–		–		–		–		–		–		–
ASTM B265-58T, grade 6, titanium alloy	1	1074		–		–		–		–		–		–		–		–
ASTM B265-58T, grade 7, titanium alloy	1	850		–		–		–		–		–		–		–		–
ASTM B301-58T, copper alloy	1	582		–		–		–		–		–		–		–		–
Astrolite, nonmetallic laminate	2	1029 1030 1052		–		–		–		–		–		–		–		–
Aurum	1	132	4	83		–		–		–		–		–		–		–
Azurite, carbonate mineral		–		–		–		–		–	8	1658		–		–		–
Baddeleytte		–		–		–		–		–		–		–		–	13	454
Bakelite		–		–		–	8	1742	8	1744	8	1746		–		–	13	1512
Balsa	2	1060		–		–		–		–		–		–		–		–
Balsa, hard pseudo	2	972 981		–		–		–		–		–		–		–		–
Balsa, waterproofed	2	1060		–		–		–		–		–		–		–		–
Balsa, x-ray protective pseudo	2	981		–		–		–		–		–		–		–		–
Banana		–		–		–		–		–		–	10	624		–		–
Barite		–		–		–		–		–	8	1702		–		–		–

Substance Name	Thermal Conductivity		Specif. Heat		Thermal Radiative Properties								Thermal Diffusivity		Viscosity		Thermal Expansion	
					Emissivity		Reflectivity		Absorptivity		Transmissiv.							
	V.	Page	V.	Page	V.	Page	V.	Page	V.	Page	V.	Page	V.	Page	V.	Page	V.	Page
Barium, Ba		–	4	13		–	7	68		–		–	10	10		–	12	21
Barium aluminum silicate, BaAl$_2$Si$_2$O$_8$		–		–		–		–		–		–		–		–	13	727
Barium boride, BaB$_6$		–		–	8	732		–		–		–		–		–		–
Barium calcium tungsten oxide, 2BaO·CaO·WO$_3$		–		–		–		--		–		–		–		–	13	577
Barium carbonate, BaCO$_3$		–	5	1109		–	8	592		–		–		–		–		–
Barium chlorides:																		
BaCl$_2$		–	5	785		–		–		–		–		–		–		–
BaCl$_2$·2H$_2$O		–	5	788		–		–		–		–		–		–		–
Barium fluoride, BaF$_2$	2	627	5	918		–	8	909		–	8	912		–		–	13	1021
Barium fluoride + calcium fluoride		–		–		–		–		–		–		–		–	13	1074
Barium oxide + silicon oxide + ΣXi, mixture	2	495		–		–		–		–		–	10	435		–		–
Barium-lead intermetallic compound, Ba$_2$Pb	1	1245		–		–		–		–		–		–		–		–
Barium metatitanate + calcium metatitanate	2	340		–		–		–		–		–		–		–		–
Barium metatitanate + magnesium zirconate, mixture	2	343		–		–		–		–		–		–		–		–
Barium metatitanate + manganese niobate, mixture	2	344		–		–		–		–		–		–		–		–
Barium nitrate, Ba(NO$_3$)$_2$		–	5	1139		–		–		–		–		–		–	13	671
Barium oxide, BaO	2	120		–		–	8	546		–		–		–		–	13	194
Barium oxide + silicon dioxide + ΣXi, mixture	2	457		–		–		–		–		–		–		–		–
Barium oxide + strontium oxide, mixture	2	337		–		–		–		–		–		–		–		–
Barium oxide + strontium oxide + ΣXi, mixture	2	460		–		–		–		–		–		–		–		–
Barium phosphate, Ba$_2$P$_2$O$_6$		–		–		–	8	608		–		–		–		–		–
Barium silicates:																		
BaSiO$_3$		–	5	1301		–		–		–		–		–		–		–
BaSi$_2$O$_5$		–	5	1307		–		–		–		–		–		–		–
Ba$_2$SiO$_4$		–	5	1304		–		–		–		–		–		–		–
Ba$_2$Si$_3$O$_8$		–	5	1310		–		–		–		–		–		–		–
Barium silicide, BaSi$_2$		–		–		–		–		–		–		–		–	13	1211
Barium stannide, Ba$_2$Sn	1	1245		–		–		–		–		–		–		–		–
Barium strontium fluoride		–		–		–	8	992		–		–		–		–		–
Barium strontium tungsten oxide, 2BaO·SrO·WO$_3$		–		–		–		–		–		–		–		–	13	578
Barium sulfate, BaSO$_4$		–	5	1176		–	8	623	8	627		–	10	413		–		–
Barium sulfate + zinc sulfide + zinc oxide, mixture		–		–		–		–		–		–	10	520		–		–
Barium sulfide, BaS		–	5	644		–		–		–		–		–		–	13	1239
Barium-tin intermetallic compound, Ba$_2$Sn	1	1246		–		–		–		–		–		–		–		–
Barium tin oxide, BaO·SnO$_2$		–		–		–		–		–		–		–		–	13	547
Barium titanium oxides:																		
BaO·TiO$_2$	2	257	5	1313		–	8	636		–	8	642		–		–	13	554
BaO·2TiO$_2$	2	260		–		–		–		–		–		–		–	13	562

18

Substance Name	Thermal Conductivity		Specif. Heat		Thermal Radiative Properties								Thermal Diffusivity		Viscosity		Thermal Expansion	
					Emissivity		Reflectivity		Absorptivity		Transmissiv.							
	V.	Page	V.	Page	V.	Page	V.	Page	V.	Page	V.	Page	V.	Page	V.	Page	V.	Page
Barium titanium oxides: (continued)																		
BaO·3TiO₂	–		–		–		–		–		–		–		–		13	562
BaO·4TiO₂	–		–		–		–		–		–		–		–		13	560
2BaO·TiO₂	–		5	1316	–		–		–		–		–		–		–	
Barium tungsten oxide, BaO·WO₃	–		–		–		8	666	–		–		–		–		–	
Barium uranium oxide, BaO·UO₃	–		5	1319	–		–		–		–		–		–		–	
Barium zirconium oxide, BaO·ZrO₂	–		5	1322	–		8	676	–		–		–		–		13	601
Barium zirconium silicate, BaZrSiO₅	–		–		–		–		8	616	–		–		–		–	
Barytes concrete	2	871	–		–		–		–		–		–		–		–	
Basalt	2	797	–		–		8	1680	–		–		–		–		–	
Basalt + dolomite	–		–		–		–		–		–		10	432	–		–	
Basalt + hornblendite	–		–		–		–		–		–		10	432	–		–	
Basalt, NTS	2	798	–		–		–		–		–		–		–		–	
Basalt, olivine	2	798	–		–		8	1688	–		–		–		–		–	
Basla, pseudo	2	1060	–		–		–		–		–		–		–		–	
Bauxite brick 20	2	902	–		–		–		–		–		–		–		–	
Bauxite brick 21	2	901	–		–		–		–		–		–		–		–	
Beef	–		–		–		–		–		–		10	625	–		–	
Beef, corned	–		–		–		–		–		–		10	625	–		–	
Beet	–		–		–		–		–		–		10	626	–		–	
Benzene	3	135	6	121	–		–		–		–		–		11	102	–	
Benzene, p-dibromo	2	986	–		–		–		–		–		–		–		–	
Benzene, p-dichloro	2	987	–		–		–		–		–		–		–		–	
Benzene, p-diiodo	2	988	–		–		–		–		–		–		–		–	
Benzene, hexadeuterated	–		6s	2	–		–		–		–		–		–		–	
Benzene–cyclohexane, mixture	–		–		–		–		–		–		–		11	350	–	
Benzene–hexane, mixture	3	387	–		–		–		–		–		–		–		–	
Benzene–n–hexane, mixture	–		–		–		–		–		–		–		11	352	–	
Benzene–octamethylcyclotetrasiloxane, mixture	–		–		–		–		–		–		–		11	354	–	
Benzenecarboxylic acid	–		6s	2	–		–		–		–		–		–		–	
1,2-Benzenediol	–		6s	83	–		–		–		–		–		–		–	
1,3-Benzenediol	–		6s	83	–		–		–		–		–		–		–	
1,4-Benzenediol	–		6s	53	–		–		–		–		–		–		–	
Benzoic acid	–		6s	2	–		–		–		–		–		–		–	
Benzol, hexadeuterated benzene	–		6s	2	–		–		–		–		–		–		–	
p-Benzoquinone	–		6s	2	–		–		–		–		–		–		–	
Benzyl acetate–m–cresol, mixture	–		–		–		–		–		–		–		11	545	–	
Benzyl alcohol	–		6s	2	–		–		–		–		–		–		–	
Benzyl benzene	–		6s	34	–		–		–		–		–		–		–	
Benzyl chloride	–		6s	22	–		–		–		–		–		–		–	
Beryl	2	800	–		–		–		–		–		–		–		–	

Substance Name	Thermal Conductivity		Specif. Heat		Thermal Radiative Properties								Thermal Diffusivity		Viscosity		Thermal Expansion	
					Emissivity		Reflectivity		Absorptivity		Transmissiv.							
	V.	Page	V.	Page	V.	Page	V.	Page	V.	Page	V.	Page	V.	Page	V.	Page	V.	Page
Beryl, India	2	801	–		–		–		–		–		–		–		–	
Berylco 25, copper alloy	–		–		–		–		–		–		–		–		12	1097
Beryllia, BeO	2	123	5	45	8	201 203 205	8	208	8	210 212	8	213	10	386	–		–	
Beryllia, brush S. P. powdered	–		–		8	202 204	–		–		–		–		–		–	
Beryllia, triangle	2	126	–		–		–		–		–		–		–		–	
Beryllium, Be	1	18	4	16	7	71 74	7	78	–		7	82	10	11	–		12	24
Beryllium, extrusion	–		–		–		7	79	–		–		–		–		–	
Beryllium, QMV, anodized	–		–		–		9	1265	–		–		–		–		–	
Beryllium alloys:																		
Be + Al	1	498	–		–		–		–		–		10	228	–		12	630
Be + Cu, anodized	–		–		9	1268	9	1270	9	1273	–		–		–		–	
Be + Fe	–		–		–		–		–		–		–		–		12	680
Be + Fe, FP-175B	–		–		–		–		–		–		–		–		12	680
Be + Mg	1	499	–		–		–		–		–		–		–		–	
Be + Al + ΣXi	–		–		–		–		–		–		–		–		12	1057
Be + Al + ΣXi, lockalloy	–		–		–		–		–		–		–		–		12	1057 1059
Be + F + ΣXi	1	929	–		–		–		–		–		–		–		–	
Be + Fe + ΣXi	–		–		–		7	1134	–		–		–		–		–	
Be + Mg + ΣXi	1	932	–		–		–		–		–		10	285	–		–	
Beryllium aluminum oxide, $BeO \cdot Al_2O_3$	–		5	1325	–		–		–		–		–		–		–	
Beryllium boride, Be_5B	–		–		8	732	–		–		–		–		–		–	
Beryllium carbide, Be_2C	2	571	–		–		–		–		–		–		–		–	
Beryllium-chromium intermetallic compound, Be_2Cr	–		–		8	1275	–		–		–		–		–		–	
Beryllium fluoride, BeF_2	–		5	921	–		–		–		–		–		–		–	
Beryllium fluoride + sodium fluoride, mixture	2	645	–		–		–		–		–		–		–		–	
Beryllium-hafnium intermetallic compound, β-$Be_{17}Hf$	–		–		–		–		–		–		–		–		12	461
Beryllium-niobium intermetallic compounds:																		
$Be_{12}Nb$	1	1248	–		8	1273	8	1280	–		–		–		–		12	464
$Be_{17}Nb_2$	1	1248	–		8	1273 1277	8	1280	–		–		–		–		–	
Beryllium oxides:																		
BeO	2	123	–		8	201 203 205	8	208	8	210 212	8	213	10	386	–		13	195
Grade, AOX	2	129	–		–		–		–		–		–		–		–	
Grade, AOX-329	2	127	–		–		–		–		–		–		–		–	
Grade, BD-98	2	125	–		–		–		–		–		–		–		–	
Grade, brush SP	2	125	–		–		–		–		–		–		–		–	
Grade, clifton metal	2	127	–		–		–		–		–		–		–		–	

Substance Name	Thermal Conduc- tivity		Specif. Heat		Thermal Radiative Properties										Thermal Diffu- sivity		Visco- sity		Thermal Expan- sion	
					Emis- sivity		Reflec- tivity		Absorp- tivity		Trans- missiv.									
	V.	Page	V.	Page	V.	Page	V.	Page	V.	Page	V.	Page	V.	Page	V.	Page	V.	Page		
Beryllium oxides: (continued)																				
Grade I	2	128	–		–		–		–		–		–		–		–			
Grade II	2	128	–		–		–		–		–		–		–		–			
Grade, UOX	2	124 127 128 129	5	45	–		–		–		–		10	386	–		–			
Norton's BeO	2	127	–		–		–		–		–		–		–		–			
Porcelain	2	124	–		–		–		–		–		–		–		–			
Refractory grade, 3008-13-3	2	125	–		–		–		–		–		–		–		–			
Triangle beryllia	2	126	–		–		–		–		–		–		–		–			
Beryllium oxide + aluminum oxide + ΣXi, mixture	2	461	–		–		–		–		–		–		–		–			
Beryllium oxide + beryllium, cermet	2 1	708 1416	5	1243 1246	–		–		–		–		–		–		13	1313		
Beryllium oxide + beryllium + molybdenum, cermet	2	711	5	1249	–		–		–		–		–		–		13	1341		
Beryllium oxide + beryllium + silicon, cermet	2	714	–		–		–		–		–		–		–		13	1341		
Beryllium oxide + beryllium tantalum compound, cermet	–		–		8	1377 1378	8	1382	–		–		–		–		–			
Beryllium oxide + magnesium oxide, mixture	2	371	–		–		–		–		–		–		–		–			
Beryllium oxide + magnesium oxide + ΣXi, mixture	2	464	–		–		–		–		–		–		–		–			
Beryllium oxide + molybdenum, cermet	–		5	1252	–		–		–		–		–		–		–			
Beryllium oxide + molybdenum beryllide, cermet	–		5	1255	–		–		–		–		–		–		–			
Beryllium oxide + niobium beryllide, cermet	–		5	1258	–		–		–		–		–		–		–			
Beryllium oxide + tantalum beryllide, cermet	–		5	1261	–		–		–		–		–		–		–			
Beryllium oxide + thorium dioxide + ΣXi, mixture	2	467	–		–		–		–		–		–		–		–			
Beryllium oxide + titanium beryllide, cermet	–		5	1264	–		–		–		–		–		–		–			
Beryllium oxide + uranium dioxide, mixture	2	347	–		–		–		–		–		–		–		–			
Beryllium oxide + zirconium beryllide, cermet	–		5	1267	–		–		–		–		–		–		–			
Beryllium oxide + zirconium dioxide + ΣXi, mixture	2	470	–		–		–		–		–		–		–		–			
Beryllium-rhenium intermetallic compound, Be_2Re	–		–		8	1275	–		–		–		–		–		–			
Beryllium-scandium intermetallic compound, $Be_{13}Sc$	–		–		8	1275	–		–		–		–		–		–			
Beryllium silicate, Be_2SiO_4	–		5	1329	–		–		–		–		–		–		–			
Beryllium sulfate, $BeSO_4$	–		5	1179	–		–		–		–		–		–		–			
Beryllium-tantalum intermetallic compounds:																				
$Be_{12}Ta$	1	1251	–		8	1273 1277	8	1280	–		–		–		–		12	467 469 470		

Substance Name	Thermal Conductivity		Specif. Heat		Thermal Radiative Properties								Thermal Diffusivity		Viscosity		Thermal Expansion	
					Emissivity		Reflectivity		Absorptivity		Transmissiv.							
	V.	Page	V.	Page	V.	Page	V.	Page	V.	Page	V.	Page	V.	Page	V.	Page	V.	Page
Beryllium-tantalum intermetallic compounds: (continued)																		
Be₁₇Ta₂	1	1251	–		8	1273 1277	8	1280	–		–		–		–		12	468 469 471
Beryllium-titanium intermetallic compound, Be₂Ti	–		–		8	1275	–		–		–		–		–		–	
Beryllium-uranium intermetallic compound, Be₁₃U	1	1254	–		–		–		–		–		–		–		12	472
Beryllium-zirconium intermetallic compounds:																		
Be₁₃Zr	1	1256	–		–		8	1280	–		–		–		–		12	475
Be₁₇Zr₂	–		–		–		8	1280	–		–		–		–		–	
Bi-sec-butyl	–		6s	33	–		–		–		–		–		–		–	
Biethylene	–		6s	5	–		–		–		–		–		–		–	
Bihexyl	–		6s	34	–		–		–		–		–		–		–	
Biisoamyl	–		6s	33	–		–		–		–		–		–		–	
Biisobutyl	–		6s	33	–		–		–		–		–		–		–	
Biisopropyl	–		6s	32	–		–		–		–		–		–		–	
Bimethyl	–		6	174	–		–		–		–		–		–		–	
Binder, 3M Kel-F 800 pigmented with:																		
SP-500 zinc oxide	–		–		–		9	317	–		–		–		–		–	
Zinc sulfide	–		–		–		9	406	–		–		–		–		–	
Binder, 3M kel-F 8213 with zinc oxide pigment	–		–		–		9	314	–		–		–		–		–	
Binder, acrylic pigmented with:																		
Aluminum, LMSC	–		–		9	3	–		9	20	–		–		–		–	
Carbon black	–		–		9	81	9	86	9	89	–		–		–		–	
Lead carbonate	–		–		–		9	139	–		–		–		–		–	
Magnesium oxide	–		–		–		9	161	–		–		–		–		–	
Strontium molybdate	–		–		9	189	9	191	9	194	–		–		–		–	
Titanium dioxide	–		–		9	211	9	223	–		–		–		–		–	
Titanium dioxide + talc	–		–		9	290	–		9	293 295	–		–		–		–	
Zinc oxide, SP-500, calcined	–		–		–		9	314	–		–		–		–		–	
Zinc sulfide	–		–		–		9	404	–		–		–		–		–	
Binder, acryloid with magnesium oxide pigment	–		–		–		9	161	–		–		–		–		–	
Binder, acryloid-A10 pigmented with:																		
Lead carbonate	–		–		–		9	138	–		–		–		–		–	
Zinc sulfide	–		–		–		9	404	–		–		–		–		–	
Binder, alkyd pigmented with:																		
Aluminum oxide	–		–		–		9	33	–		–		–		–		–	
Lead carbonate	–		–		–		9	139	–		–		–		–		–	
Magnesium oxide	–		–		–		9	161	–		–		–		–		–	
Titanium dioxide	–		–		9	211 214	9	221 250	9	256	–		–		–		–	

Substance Name	Thermal Conductivity		Specif. Heat		Thermal Radiative Properties								Thermal Diffusivity		Viscosity		Thermal Expansion	
					Emissivity		Reflectivity		Absorptivity		Transmissiv.							
	V.	Page	V.	Page	V.	Page	V.	Page	V.	Page	V.	Page	V.	Page	V.	Page	V.	Page
Binder, alkyd pigmented with: (continued)																		
Titanium dioxide + calcium sulfate	-		-			-	9	289	-		-		-		-		-	
Titanox AMO	-		-			-	9	222	-		-		-		-		-	
Zinc sulfide + clay	-		-			-		-	9	416	-		-		-		-	
Binder, alkyd-melamine pigmented with:																		
Aluminum oxide	-		-			-	9	33	-		-		-		-		-	
$CaSO_4$ + TiO_2	-		-			-	9	78	-		-		-		-		-	
Lead carbonate	-		-			-	9	139	-		-		-		-		-	
Magnesium oxide	-		-			-	9	161	-		-		-		-		-	
Titanium dioxide	-		-			-	9	221	-		-		-		-		-	
Titanox C-50	-		-			-	9	78 289	-		-		-		-		-	
Titanox RC	-		-			-	9	78 289	-		-		-		-		-	
Binder, aluminum phosphate pigmented with:																		
Barium titanate	-		-		9	63	-		-		-		-		-		-	
Calcium titanate	-		-		9	79 80	-		-		-		-		-		-	
Cr-Co-Ni spinel	-		-		9	186	-		-		-		-		-		-	
FCE-11	-		-		9	63	-		-		-		-		-		-	
Iron titanate	-		-		9	123	-		-		-		-		-		-	
Iron titanate + alumina	-		-		9	123	-		-		-		-		-		-	
$NiO \cdot Cr_2O_3$ spinel + SiO_2	-		-		9	186	-		-		-		-		-		-	
Silicon carbide	-		-		9	174	-		-		-		-		-		-	
SiC + SiO_2	-		-		9	174	-		-		-		-		-		-	
Strontium titanate	-		-		9	197	-		-		-		-		-		-	
Tin oxide	-		-		9	201	-		9	205	-		-		-		-	
Ultrox, $ZrSiO_4$	-		-			-		-	9	435	-		-		-		-	
Zirconium oxide	-		-			-	9	425	-		-		-		-		-	
Zirconium silicate	-		-			-		-	9	435	-		-		-		-	
Binder, barium beryllium silicate with cerium dioxide pigment	-		-		9	445 448	-		-		-		-		-		-	
Binder, barium borosilicate frit with chromium oxide pigment	-		-		9	455 459	-		-		-		-		-		-	
Binder, base glaze No. 1 pigmented with:																		
Cerium dioxide	-		-		9	445	-		-		-		-		-		-	
Chromium oxide + cobalt oxide	-		-		9	455	-		-		-		-		-		-	
Cobalt oxide + chromium oxide	-		-		9	464	-		-		-		-		-		-	
Binder, base glaze no. 2 pigmented with:																		
Chromium oxide + cobalt oxide	-		-		9	456	-		-		-		-		-		-	
Cobalt oxide + chromium oxide	-		-		9	464	-		-		-		-		-		-	
Cobalt oxide + manganese oxide	-		-		9	464	-		-		-		-		-		-	
Manganese oxide + cobalt oxide	-		-		9	468	-		-		-		-		-		-	

Substance Name	Thermal Conductivity		Specif. Heat		Thermal Radiative Properties								Thermal Diffusivity		Viscosity		Thermal Expansion	
					Emissivity		Reflectivity		Absorptivity		Transmissiv.							
	V.	Page	V.	Page	V.	Page	V.	Page	V.	Page	V.	Page	V.	Page	V.	Page	V.	Page
Binder, base glaze no. 3 pigmented with:																		
Cerium dioxide + cobalt oxide	–		–		9	445	–		–		–		–		–		–	
Chromium oxide + iron oxide	–		–		9	456	–		–		–		–		–		–	
Cobalt oxide	–		–		9	464	–		–		–		–		–		–	
Cobalt oxide + nickel oxide	–		–		9	464	–		–		–		–		–		–	
Iron oxide + chromium oxide	–		–		9	466	–		–		–		–		–		–	
Iron oxide + manganese oxide	–		–		9	466	–		–		–		–		–		–	
Iron oxide + nickel oxide	–		–		9	466	–		–		–		–		–		–	
Manganese oxide + cobalt oxide	–		–		9	468	–		–		–		–		–		–	
Manganese oxide + iron oxide	–		–		9	468	–		–		–		–		–		–	
Nickel oxide + cobalt oxide	–		–		9	469	–		–		–		–		–		–	
Nickel oxide + iron oxide	–		–		9	469	–		–		–		–		–		–	
Binder, carboxy-methyl-cellulose with barium sulfate pigment	–		–		–		9	62	–		–		–		–		–	
Binder, cellulose nitrate with titanium dioxide pigment	–		–		–		–		9	256	–		–		–		–	
Binder, copolymer acetone with zinc oxide pigment	–		–		–		9	314	–		–		–		–		–	
Binder, decoret with carbon black pigment	–		–		–		9	86	–		–		–		–		–	
Binder, decoret with lampblack pigment	–		–		–		9	86	–		–		–		–		–	
Binder, Dow Corning 20 with silicon carbide pigment	–		–		9	174	–		–		–		–		–		–	
Binder, Dow Corning 806 A pigmented with:																		
Boron nitride	–		–		–		9	68	–		–		–		–		–	
Boron nitride + diatomaceous earth	–		–		–		9	70	–		–		–		–		–	
Diatomaceous earth	–		–		–		9	113	–		–		–		–		–	
Magnesium oxide	–		–		–		9	161	–		–		–		–		–	
Magnesium oxide + diatomaceous earth	–		–		–		9	165	–		–		–		–		–	
SiC + TiC	–		–		9	174	–		–		–		–		–		–	
Titanium dioxide	–		–		9	211	9	222	–		–		–		–		–	
Zinc oxide	–		–		–		9	317	–		–		–		–		–	
Binder, Dow Corning 807 with lampblack pigment	–		–		9	82	9	86	–		–		–		–		–	
Binder, Dow Corning Q90016 with zinc oxide pigment	–		–		–		9	308 321	9	375	–		–		–		–	
Binder, Dow Corning Q90090 with zinc oxide pigment	–		–		–		–		9	362	–		–		–		–	
Binder, Du Pont RC-7007 pigmented with:																		
Aluminum oxide	–		–		–		9	33	–		–		–		–		–	
$CaSO_4 + TiO_2$	–		–		–		9	78	–		–		–		–		–	
Du Pont R-100 pigment	–		–		–		9	223	–		–		–		–		–	
Du Pont R-510 pigment	–		–		–		9	221	–		–		–		–		–	
Magnesium oxide	–		–		–		9	161	–		–		–		–		–	

Substance Name	Thermal Conductivity V.	Page	Specif. Heat V.	Page	Thermal Radiative Properties Emissivity V.	Page	Reflectivity V.	Page	Absorptivity V.	Page	Transmissiv. V.	Page	Thermal Diffusivity V.	Page	Viscosity V.	Page	Thermal Expansion V.	Page
Binder, Du Pont RC-7007 pigmented with: (continued)																		
Titanium dioxide	-		-			-	9	221	-		-		-		-		-	
Titanox C-50	-		-			-	9	78 289	-		-		-		-		-	
Titanox RC	-		-			-	9	78 289	-		-		-		-		-	
Binder, Du Pont viton B with zinc oxide pigment	-		-			-	9	314	-		-		-		-		-	
Binder, epoxide with titanium dioxide pigment	-		-		9	211		-	-		-		-		-		-	
Binder, epoxy pigmented with:																		
Carbon black	-		-		9	82 84	9	86	9	89	-		-		-		-	
Lampblack	-		-		9	82 84	9	86	9	89	-		-		-		-	
Titanium dioxide	-		-		9	211		-	9	252 255 263 281	-		-		-		-	
Binder, ethylcellulose + Dow 7 pigmented with:																		
Magnesium carbonate	-		-			-	9	156	-		-		-		-		-	
Magnesium oxide	-		-			-	9	158	-		-		-		-		-	
Binder, formaldehyde with lead carbonate pigment	-		-			-	9	138	-		-		-		-		-	
Binder, G.E. RTV-602 silicone resin pigmented with:																		
China clay	-		-		9	95	9	97	9	98 100	-		-		-		-	
Rutile TiO$_2$ Du Pont R-960	-		-			-	9	228	9	273	-		-		-		-	
Titanium oxide	-		-			-	9	220 244	9	256 263	-		-		-		-	
Zinc oxide	-		-			-	9	314	9	373 392	-		-		-		-	
Zinc oxide, S-13G	-		-		9	304	9	322 355	9	382 392	-		-		-		-	
Zinc oxide, S-13H	-		-			-	9	323	-		-		-		-		-	
Binder, G.E. SE 551 methyl-phenyl silicone with zinc oxide pigment	-		-			-	9	317	-		-		-		-		-	
Binder, G.E. SR-122 with magnesium oxide pigment	-		-			-		-	9	163	-		-		-		-	
Binder, gelatin with silver chloride pigment	-		-			-	9	181	-		9	183	-		-		-	
Binder, lacquer with aluminum pigment	-		-		9	6		-	-		-		-		-		-	
Binder, leonite 201-S pigmented with:																		
Magnesium oxide	-		-			-	9	161	-		-		-		-		-	
Zinc oxide	-		-			-	9	314	-		-		-		-		-	
Zinc sulfide	-		-			-	9	406	-		-		-		-		-	
Binder, LTV-602 with SP-500 zinc oxide pigment	-		-			-	9	322	-		-		-		-		-	
Binder, LTV-602 with Ti pure R-900-1 pigment	-		-			-	9	220	-		-		-		-		-	

Substance Name	Thermal Conduc- tivity		Specif. Heat		Thermal Radiative Properties								Thermal Diffu- sivity		Visco- sity		Thermal Expan- sion	
					Emis- sivity		Reflec- tivity		Absorp- tivity		Trans- missiv.							
	V.	Page	V.	Page	V.	Page	V.	Page	V.	Page	V.	Page	V.	Page	V.	Page	V.	Page
Binder, LTV-602 with titanox RA-10 pigment	–		–			–	9	220	–		–		–		–		–	
Binder, ludox with SiC + talc pigment	–		–		9	176		–	–		–		–		–		–	
Binder, methylphenol silicone with anatase TiO₂ pigment	–		–			–	9	244	–		–		–		–		–	
Binder, NBS frit No. 332 pigmented with:																		
Cerium dioxide + magnesium oxide	–		–			–	9	452	–		–		–		–		–	
Cerium dioxide + tin oxide	–		–		9	445	9	452	–		–		–		–		–	
Cerium dioxide + zirconium oxide	–		–			–	9	452	–		–		–		–		–	
Chromium oxide	–		–		9	455	9	463	–		–		–		–		–	
Chromium oxide + black stain	–		–		9	455	9	463	–		–		–		–		–	
CoO·Cr₂O₃ spinel	–		–		9	472	9	475	–		–		–		–		–	
CoO·Fe₂O₃ spinel	–		–		9	472	9	475	–		–		–		–		–	
CoO·Mn₂O₃ spinel	–		–		9	472	9	475	–		–		–		–		–	
Fe₂O₃ + CoO + Cr₂O₃	–		–		9	466	9	467	–		–		–		–		–	
NiO·Cr₂O₃ spinel	–		–		9	472	9	475	–		–		–		–		–	
NiO·Fe₂O₃ spinel	–		–		9	472	9	475	–		–		–		–		–	
Tin oxide + cerium dioxide	–		–		9	477	9	478	–		–		–		–		–	
Zirconium oxide + cerium dioxide	–		–		9	479		–	–		–		–		–		–	
Binder, necoloidine pigmented with:																		
Titanium hydride	–		–		9	209		–	–		–		–		–		–	
Zirconium carbide	–		–		9	421		–	–		–		–		–		–	
Zirconium hydride + iron oxide	–		–		9	422		–	–		–		–		–		–	
Zirconium hydride + zirconia	–		–		9	422		–	–		–		–		–		–	
Binder, nitrocellulose pigmented with:																		
Aluminum	–		–		9	3		–	–		–		–		–		–	
Carbon black	–		–			–	9	86	–		–		–		–		–	
Lead carbonate	–		–			–	9	139	–		–		–		–		–	
Titanium dioxide	–		–		9	211	9	221	–		–		–		–		–	
Titanium pyrophosphate	–		–		9	298		–	–		–		–		–		–	
Zirconium oxide	–		–		9	423		–	–		–		–		–		–	
Binder, oil with aluminum pigment	–		–			–	9	9	–		–		–		–		–	
Binder, oil with leafing aluminum pigment	–		–			–	9	9	–		–		–		–		–	
Binder, Owens-Illinois 650 with SP-500 zinc oxide pigment	–		–			–	9	322	9	377 393	–		–		–		–	
Binder, phosphate with potassium titanate pigment	–		–		9	172		–	–		–		–		–		–	
Binder, phosphoric acid cement with aluminum oxide pigment	–		–		9	30		–	–		–		–		–		–	
Binder, phthalic alkyd pigmented with:																		
Carbon black	–		–			–		–	9	89	–		–		–		–	
Carbon + aluminum	–		–		9	91		–	9	93	–		–		–		–	
Binder, polyvinyl alcohol with barium sulfate pigment	–		–			–	9	62	–		–		–		–		–	

Substance Name	Thermal Conduc-tivity		Specif. Heat		Thermal Radiative Properties								Thermal Diffu-sivity		Visco-sity		Thermal Expan-sion	
					Emis-sivity		Reflec-tivity		Absorp-tivity		Trans-missiv.							
	V.	Page	V.	Page	V.	Page	V.	Page	V.	Page	V.	Page	V.	Page	V.	Page	V.	Page
Binder, polyvinyl butyral pigmented with:																		
Titanium dioxide	–		–			–		–	9	267	–		–		–		–	
Zinc sulfide	–		–			–	9	304	9	415	–		–		–		–	
Binder, PS-7 potassium silicate pigmented with:																		
Alucer MC Al_2O_3	–		–			–		–	9	41	–		–		–		–	
Aluminum oxide	–		–		9	28	9	33 39	9	41 42	–		–		–		–	
Al_2O_3 + TiO_2 + ZnO	–		–		9	28	9	34	9	42	–		–		–		–	
Aluminum phosphate	–		–			–		–	9	435	–		–		–		–	
Aluminum silicate	–		–			–	9	47	9	49	–		–		–		–	
Antimony oxide	–		–			–	9	51	9	54 56 57 60	–		–		–		–	
Boron nitride	–		–		9	66		–		–	–		–		–		–	
Cabot RF-1 TiO_2	–		–		9	214		–		–	–		–		–		–	
Calcium metasilicate	–		–			–	9	76		–	–		–		–		–	
Cr-Co-Ni spinel	–		–			–		–	9	187	–		–		–		–	
Diatomaceous earth	–		–			–	9	113	9	116	–		–		–		–	
Dicalite WB-5	–		–			–	9	113	9	179	–		–		–		–	
E-P730 zinc oxide	–		–			–	9	316		–	–		–		–		–	
Lanthanum oxide	–		–			–	9	127	9	132	–		–		–		–	
Lithafrax	–		–			–	9	144		–	–		–		–		–	
Lithium aluminum silicate	–		–		9	143	9	144		–	–		–		–		–	
Magnesium aluminate spinel	–		–			–		–	9	187	–		–		–		–	
Magnesium silicate	–		–			–	9	168	9	169	–		–		–		–	
Molochite no. 6	–		–			–	9	47	9	49	–		–		–		–	
Molochite SF	–		–			–	9	47	9	49	–		–		–		–	
Silicon dioxide	–		–			–	9	178	9	179	–		–		–		–	
SP-500 zinc oxide pigment	–		–		9	304	9	314	9	360 365 370 394	–		–		–		–	
Strontium molybdate	–		–		9	188	9	191	9	194	–		–		–		–	
Superpax $ZrSiO_4$ pigment	–		–			–	8	432	8	437	–		–		–		–	
Tin oxide	–		–		9	201	9	204	9	205 207	–		–		–		–	
Titanium dioxide	–		–		9	214	9	226	9	274 281	–		–		–		–	
Titanium dioxide + aluminum oxide, mixture	–		–			–	9	285 287		–	–		–		–		–	
XX254 ZnO pigment	–		–			–	9	316		–	–		–		–		–	
Zinc oxide	–		–		9	302 304	9	306 314 355	9	360 362 365 373 392	–		–		–		–	
Zinc sulfide	–		–			–	9	405	9	418	–		–		–		–	

Substance Name	Thermal Conductivity		Specif. Heat		Thermal Radiative Properties										Thermal Diffusivity		Viscosity		Thermal Expansion	
					Emissivity		Reflectivity		Absorptivity		Transmissiv.									
	V.	Page	V.	Page	V.	Page	V.	Page	V.	Page	V.	Page	V.	Page	V.	Page	V.	Page	V.	Page
Binder, PS-7 potassium silicate pigmented with: (continued)																				
Zinc titanate	–		–			–		–	9	420	–		–			–		–		–
Zirconium oxide	–		–			–	9	425	9	428	–		–			–		–		–
Zirconium silicate	–		–			–	9	432	9	435 437 441	–		–			–		–		–
Binder, R-44 acrylic with strontium molybdate pigment	–		–			–	9	191	9	194	–		–			–		–		–
Binder, resin with leafing gold pigment	–		–		9	25		–		–	–		–			–		–		–
Binder, RTV-11 polymethyl siloxane with zinc oxide pigment	–		–			–	9	317		–	–		–			–		–		–
Binder, silica pigmented with:																				
Silicon carbide	–		–		9	174		–		–	–		–			–		–		–
Titanium carbide	–		–		9	208		–		–	–		–			–		–		–
Zirconium oxide	–		–			–	9	425		–	–		–			–		–		–
Binder, silicate with tin oxide pigment	–		–			–		–	9	205 207	–		–			–		–		–
Binder, silicone-alkyd with zinc sulfide + clay pigment	–		–			–		–	9	416	–		–			–		–		–
Binder, silicone alkyd epoxide with titanium dioxide pigment	–		–		9	212	9	221 250	9	275	–		–			–		–		–
Binder, silicone pigmented with:																				
Aluminum	–		–		9	3 6	9	13 17	9	21	–		–			–		–		–
Aluminum + carbon	–		–			–		–	9	24	–		–			–		–		–
Antimony oxide	–		–			–	9	51		–	–		–			–		–		–
Barium titanate	–		–		9	63		–	9	64	–		–			–		–		–
Boron nitride	–		–			–	9	68		–	–		–			–		–		–
Boron nitride + diatomaceous earth	–		–			–	9	70		–	–		–			–		–		–
Calcium carbonate	–		–			–	9	73		–	–		–			–		–		–
Carbon black	–		–		9	81 82		–	9	89	–		–			–		–		–
China clay	–		–		9	95	9	97	9	98 100	–		–			–		–		–
Clay + TiO$_2$	–		–		9	105		–	9	107	–		–			–		–		–
Diatomaceous earth	–		–			–	9	113		–	–		–			–		–		–
Iron oxide	–		–		9	117 119 121		–		–	–		–			–		–		–
Lampblack	–		–		9	82	9	86		–	–		–			–		–		–
Lead carbonate	–		–			–	9	138		–	–		–			–		–		–
Leafing aluminum	–		–		9	3 7	9	13 17	9	21	–		–			–		–		–
Leafing aluminum + carbon	–		–			–		–	9	24	–		–			–		–		–
Magnesium oxide	–		–			–	9	161		–	–		–			–		–		–
Magnesium oxide + diatomaceous earth	–		–			–	9	165		–	–		–			–		–		–
Micro-cell C, diatomaceous earth	–		–			–	9	113		–	–		–			–		–		–

Substance Name	Thermal Conduc- tivity		Specif. Heat		Thermal Radiative Properties								Thermal Diffu- sivity		Visco- sity		Thermal Expan- sion	
					Emis- sivity		Reflec- tivity		Absorp- tivity		Trans- missiv.								
	V.	Page	V.	Page	V.	Page	V.	Page	V.	Page	V.	Page	V.	Page	V.	Page	V.	Page
Binder, silicone pigmented with: (continued)																		
Silicon dioxide		–		–		–	9	178		–		–		–		–		–
SP-500 zinc oxide pigment		–		–	9	304	9	306 308 314	9	370 394		–		–		–		–
Strontium zirconate		–		–	9	198		–	9	199		–		–		–		–
Superlith XXXN pigment		–		–		–	8	405	8	415		–		–		–		–
Titanium oxide		–		–	9	211	9	220 244	9	252 255 263 281		–		–		–		–
Titanium oxide, Thermatrol ZA-100		–		–		–	9	224		–		–		–		–		–
Titanox A-WD, TiO₂		–		–		–	9	222		–		–		–		–		–
Titanox AMO, anatase TiO₂		–		–		–	9	222		–		–		–		–		–
Zinc oxide		–		–	9	302 304	9	308 314 355	9	362 371 392		–		–		–		–
Zinc oxide, B-1060		–		–	9	304		–	9	382 393		–		–		–		–
Zinc oxide, S-13		–		–		–	9	318 355	9	371 392		–		–		–		–
Zinc sulfide		–		–	9	401	9	404	9	412 415		–		–		–		–
Zirconium silicate		–		–	9	430		–	9	437		–		–		–		–
Binder, silicone ZW 40 with zinc sulfide pigment		–		–	9	401		–	9	415		–		–		–		–
Binder, silicone ZW 60 with zinc sulfide pigment		–		–	9	401		–	9	415		–		–		–		–
Binder, siloxane with titanium dioxide pigment		–		–		–	9	220	9	255 264		–		–		–		–
Binder, siloxane with zinc oxide pigment		–		–		–	9	316	9	362 392		–		–		–		–
Binder, sodium silicate pigmented with:																		
Aquablack, B		–		–	9	574		–	9	575		–		–		–		–
Calcium fluoride		–		–		–		–	9	75		–		–		–		–
Lithafrax		–		–	9	142		–	9	147 150		–		–		–		–
Lithium aluminum silicate		–		–	9	142		–	9	147 150		–		–		–		–
Lithium fluoride		–		–		–		–	9	155		–		–		–		–
Potassium aluminum silicate		–		–		–		–	9	171		–		–		–		–
Sodium aluminum silicate		–		–		–		–	9	185		–		–		–		–
Spodumene		–		–		–		–	9	150 153		–		–		–		–
Titanium dioxide		–		–		–		–	9	255		–		–		–		–
Zinc oxide		–		–		–		–	9	362		–		–		–		–
Zinc sulfide		–		–		–		–	9	412		–		–		–		–
Zirconium		–		–	9	26		–		–		–		–		–		–
Zirconium silicate		–		–	9	430		–	9	437		–		–		–		–

Substance Name	Thermal Conductivity		Specif. Heat		Thermal Radiative Properties								Thermal Diffusivity		Viscosity		Thermal Expansion	
					Emissivity		Reflectivity		Absorptivity		Transmissiv.							
	V.	Page	V.	Page	V.	Page	V.	Page	V.	Page	V.	Page	V.	Page	V.	Page	V.	Page
Binder, synar pigmented with:																		
Boron carbide		–		–	9	65		–		–		–		–		–		–
$Cr_2O_3 + Fe_3O_4 + NiO$		–		–	9	103		–		–		–		–		–		–
Silicon carbide		–		–	9	174		–		–		–		–		–		–
$SiC + UO_2$		–		–	9	176		–		–		–		–		–		–
Binder, teflon pigmented with:																		
Zinc oxide		–		–		–	9	317		–		–		–		–		–
Zirconium oxide		–		–		–	9	425		–		–		–		–		–
Binder, turpentine with carbon black pigment		–		–		–	9	86		–		–		–		–		–
Binder, turpentine with lampblack pigment		–		–		–	9	86		–		–		–		–		–
Binder, viton with zinc oxide pigment		–		–		–	9	317		–		–		–		–		–
Binder, viton B copolymer with SP-500 zinc oxide pigment		–		–		–	9	314		–		–		–		–		–
Binder, xylol pigmented with:																		
Acetylene black		–		–	9	81		–		–		–		–		–		–
Carbon black		–		–	9	81		–		–		–		–		–		–
1,2-Binitrobenzene		–	6s	34		–		–		–		–		–		–		–
1,3-Binitrobenzene		–	6s	34		–		–		–		–		–		–		–
1,4-Binitrobenzene		–	6s	34		–		–		–		–		–		–		–
1,2-Binitrobenzol		–	6s	34		–		–		–		–		–		–		–
1,3-Binitrobenzol		–	6s	34		–		–		–		–		–		–		–
1,4-Binitrobenzol		–	6s	34		–		–		–		–		–		–		–
Bioctyl		–	6s	43		–		–		–		–		–		–		–
Biphenyl	2	989		–		–		–		–		–		–		–		–
Biphenyl + o-,m-,p-terphenyl + higher phenyls, santowax R	2	1005		–		–		–		–		–		–		–		–
Bismuth alloys:																		
Bismuth, Bi	1	25	4	21		–		–		–	7	85 88	10	12		–	12	33
Bismuth alloys:																		
Bi + Cd	1	505		–		–		–		–		–		–		–		–
Bi + Pb	1	508	4	291		–		–		–		–		–		–	12	681
Bi + Pb, eutectic alloy	1	509		–		–		–		–		–		–		–		–
Bi + Sb	1	502		–		–		–		–		–		–		–	12	673
Bi + Sn	1	511		–		–		–	7	899		–		–		–	12	684
Bi + Sn, Hutchins alloy	1	512		–		–		–		–		–		–		–		–
Bi + Cd + ΣXi	1	935		–		–		–		–		–		–		–		–

Substance Name	Thermal Conductivity		Specif. Heat		Thermal Radiative Properties								Thermal Diffusivity		Viscosity		Thermal Expansion	
					Emissivity		Reflectivity		Absorptivity		Transmissiv.							
	V.	Page	V.	Page	V.	Page	V.	Page	V.	Page	V.	Page	V.	Page	V.	Page	V.	Page
Bismuth alloys: (continued)																		
Bi + Pb + ΣXi	1	938	–		–		–		–		–		–		–		–	
Bi + Pb + ΣXi, Lipowitz alloy	1	939	–		–		–		–		–		–		–		–	
Bi + Pb + ΣXi, Rose metal	1	939	–		–		–		–		–		–		–		–	
Bi + Pb + ΣXi, Woods metal	1	939	–		–		–		–		–		–		–		–	
Bismuth glance, Bi_2Te_3	–		5	717	–		–		–		–		–		–		–	
Bismuth iodide, BiI_3	–		–		–		8	1027	–		–		–		–		–	
Bismuth oxide, Bi_2O_3	–		5	48	–		–		–		–		–		–		–	
Bismuth-platinum intermetallic compounds:																		
BiPt	–		–		–		–		–		–		–		–		12	479 480 482
Bi_2Pt	–		–		–		–		–		–		–		–		12	478 480 481
Bismuth selenide, Bi_2Se_3	–		–		–		8	1130	–		–		–		–		–	
Bismuth selenide + bismuth telluride, mixture	1	1393	–		–		–		–		–		–		–		–	
Bismuth stannate, $Bi_2(SnO_3)_3$	2	261	–		–		–		–		–		–		–		–	
Bismuth sulfide, Bi_2S_3	–		5	647	–		–		–		–		–		–		–	
Bismuth telluride, Bi_2Te_3	1	1257	5	717	–		8	1238	–		–		10	456	–		13	1270
Bismuth tellurium selenide	–		–		–		8	1130	–		–		–		–		–	
Bismuth telluride + tellurium, mixture	1	1415	–		–		–		–		–		–		–		–	
Bismuth titanium oxide	–		–		–		–		–		8	644	–		–		–	
Bisphenol-A	–		–		–		–		–		–		–		–		13	1405
Bitter spar, dolomite	2	810	5	1115	–		–		–		–		–		–		–	
Bitumen	2	1155	–		–		–		–		–		–		–		–	
Bitumin concrete	2	863	–		–		–		–		–		–		–		–	
Bivinyl	–		6s	5	–		–		–		–		–		–		–	
Blanc fixe	–		–		–		–		–		–		10	413	–		–	
Bone char	2	1156	–		–		–		–		–		–		–		–	
Boric acid + titanium boride, mixture	–		–		8	1468 1469	8	1471	–		–		–		–		–	
Boric acid + titanium boride + titanium oxide, powder	–		–		8	1515 1516	8	1518	–		–		–		–		–	
Boric oxide glass	–		–		–		–		–		–		–		–		13	1352
Boron, B	1	41	4	25	–		–		–		–		10	16	–		13	12
Boron/Avco 5505, composite	–		–		–		–		–		–		–		–		13	1533
Boron carbide, B_4C	2	572	5	402	8	852	8	855	–		–		10	461	–		13	840
Boron carbide + aluminum, cermet	2	717	–		–		–		–		–		–		–		–	
Boron carbide + molybdanum oxide powders	–		–		–		8	1465	–		–		–		–		–	
Boron carbide + sodium silicate, mixture	2	541	–		–		–		–		–		–		–		–	
Boron fluoride oxide, trimeric	–		6s	2	–		–		–		–		–		–		–	
Boron nitride, BN	2	656	5	1078	8	1037 1040 1042	8	1047	–		8	1054	–		–		13	1131

Substance Name	Thermal Conduc- tivity		Specif. Heat		Thermal Radiative Properties										Thermal Diffu- sivity		Visco- sity		Thermal Expan- sion	
					Emis- sivity		Reflec- tivity		Absorp- tivity		Trans- missiv.									
	V.	Page	V.	Page	V.	Page	V.	Page	V.	Page	V.	Page	V.	Page	V.	Page	V.	Page	V.	Page
Boron nitride, boralloy	2	656	5	1078		–		–		–		–		–		–		–		
Boron nitride + boron oxide + ΣXi, mixture		–	5	1270		–		–		–		–		–		–		–		
Boron nitride + carbon, mixture		–	5	1273		–		–		–		–		–		–		–		
Boron oxide, B_2O_3	2	138	5	51		–		–		–		–		–		–		–		
Boron oxide + silicon oxide, mixture	2	498		–		–		–		–		–	10	436 437		–		–		
Boron oxyfluoride, trimeric		–	6s	2		–		–		–		–		–		–		–		
Boron phosphate		–		–		–	8	608		–		–		–		–		–		
Boron phosphide, BP		–		–	8	1105	8	1107		–		–		–		–	13	1165		
Boron silicides:																				
SIB_4	1	1262		–	8	1134 1136	8	1138		–		–		–		–		–		
SIB_6	1	1262		–	8	1134 1136	8	1138		–		–		–		–		–		
Boron + titanium boride powder		–		–		–	8	1447		–		–		–		–		–		
Boron tribromide, BBr_3		–	6s	2		–		–		–		–		–		–		–		
Boron trichloride, BCl_3		–	6s	2		–		–		–		–		–		–		–		
Boron trifluoride, BF_3	3	99	6	67		–		–		–		–		–	11	74		–		
Borosilicate, powder	2	1040		–		–		–		–		–		–		–		–		
Boxwood	2	1061		–		–		–		–		–		–		–		–		
Brain tissue		–		–		–		–		–		–	10	627		–		–		
Brain tissue, cat		–		–		–		–		–		–	10	628		–		–		
Brass		–	4	346	7	912 915		–	7	925		–	10	239		–	12	1116		
Brass, 70/30	1	590		–		–		–		–		–		–		–	12	1116		
Brass, alpha		–		–		–	7	921		–		–		–		–		–		
Brass, B. S. 249	1	981		–		–		–		–		–		–		–		–		
Brass, cast	1	980		–		–		–		–		–		–		–		–		
Brass, common		–		–		–		–		–		–		–		–	12	798		
Brass, German, red	1	981		–		–		–		–		–		–		–		–		
Brass, high	1	981 982		–		–		–		–		–		–		–		–		
Brass, high tensile	1	980		–		–		–		–		–		–		–		–		
Brass, leaded free cutting	1	981		–		–		–		–		–		–		–		–		
Brass, MS 58	1	980		–		–		–		–		–		–		–		–		
Brass, MS 76/22/2	1	980		–		–		–		–		–		–		–		–		
Brass, R		–		–		–	7	921		–		–		–		–		–		
Brass, red	1	591		–		–		–		–		–		–		–	12	1116		
Brass, rolled	1	980		–		–		–		–		–		–		–		–		
Brass, yellow	1	981 982		–	7	912		–	7	923		–	10	239		–		–		
Brass, yellow ASTM B16		–		–		–		–		–		–		–		–	12	1116		
Brazil beryl	2	801		–		–		–		–		–		–		–		–		
Brazil topaz	2	252		–		–		–		–		–		–		–		–		
Brazil tourmaline	2	855		–		–		–		–		–		–		–		–		

32

Substance Name	Thermal Conductivity		Specif. Heat		Thermal Radiative Properties								Thermal Diffusivity		Viscosity		Thermal Expansion	
					Emissivity		Reflectivity		Absorptivity		Transmissiv.							
	V.	Page	V.	Page	V.	Page	V.	Page	V.	Page	V.	Page	V.	Page	V.	Page	V.	Page
Brick, bauxite	2	901 902	–		–		–		–		–		–		–		–	
Brick, calcined, Sil-O-Cel	2	896	–		–		–		–		–		–		–		–	
Brick, Carbofrax	2	897	–		–		–		–		–		–		–		–	
Brick, Carbofrax carborundum	2	895	–		–		–		–		–		–		–		–	
Brick, carbon	2	890 896	–		–		–		–		–		–		–		–	
Brick, carslat carborundum	2	895	–		–		–		–		–		–		–		–	
Brick, cement porous	2	890	–		–		–		–		–		–		–		–	
Brick, ceramic	2	890	–		–		–		–		–		–		–		–	
Brick, chamotte	2	890	–		–		–		–		–		–		–		–	
Brick, chrome	2	454 897 898	–		–		–		–		–		–		–		–	
Brick, chrome fire	2	897	–		–		–		–		–		–		–		–	
Brick, chrome magnesite	2	890	–		–		–		–		–		–		–		–	
Brick, chromite	2	473 899	–		–		–		–		–		–		–		–	
Brick, chromomagnesite	2	481	–		–		–		–		–		–		–		–	
Brick, common	2	488 489 900 901	–		–		–		–		–		–		–		–	
Brick, corundum	2	454 905	–		–		–		–		–		–		–		–	
Brick, cupola		–	–		–		–		–		–		10	570	–		–	
Brick, dense	2	443 904	–		–		–		–		–		–		–		–	
Brick, dense fireclay	2	403	–		–		–		–		–		–		–		–	
Brick, diatomaceous	2	890 891	–		–		–		–		–		–		–		–	
Brick, diatomaceous insulating	2	906 907	–		–		–		–		–		–		–		–	
Brick, dinas	2	891	–		–		–		–		–		–		–		–	
Brick, Egyptian fire clay	2	491 901	–		–		–		–		–		–		–		–	
Brick, fire	2	491 891 895 902 903	–		–		–		–		–		10	570	–		–	
Brick, fire, Missouri	2	905	–		–		–		–		–		–		–		–	
Brick, fire clay	2	403 404 490 491 896 901 903	–		–		–		–		–		–		–		–	
Brick, fire clay aluminous	2	900	–		–		–		–		–		–		–		–	
Brick, fire clay dense	2	903	–		–		–		–		–		–		–		–	
Brick, fire clay superduty	2	890	–		–		–		–		–		–		–		–	
Brick, fused alumina	2	897	–		–		–		–		–		–		–		–	
Brick, Georgia fire	2	896	–		–		–		–		–		–		–		–	

Substance Name	Thermal Conductivity		Specif. Heat		Thermal Radiative Properties											Thermal Diffusivity		Viscosity		Thermal Expansion	
					Emissivity		Reflectivity		Absorptivity		Transmissiv.										
	V.	Page	V.	Page	V.	Page	V.	Page	V.	Page	V.	Page		V.	Page	V.	Page	V.	Page	V.	Page
Brick, hand-burned face	2	891	–		–		–		–		–			–		–		–			
Brick, high temp. insulating	2	891	–		–		–		–		–			–		–		–			
Brick, high temp. insulating blast furnace	2	899	–		–		–		–		–			–		–		–			
Brick, Hytex building	2	896	–		–		–		–		–			–		–		–			
Brick, insulating	2	443 891 904	–		–		–		–		–			–		–		–			
Brick, insulating fire	2	891	–		–		–		–		–			–		–		–			
Brick, Italy, porous fire	2	895	–		–		–		–		–			–		–		–			
Brick, kaolin fire	2	404 405 904	–		–		–		–		–			–		–		–			
Brick, kaolin insulating refractory	2	895	–		–		–		–		–			–		–		–			
Brick, ladle		–		–		–		–		–		–		10	570		–		–		
Brick, light weight	2	488 489 892 899 900	–		–		–		–		–			–		–		–			
Brick, lime sand	2	892	–		–		–		–		–			–		–		–			
Brick, magnesia	2	485 897 898 899	–		–		–		–		–			–		–		–			
Brick, magnesite	2	478 483 892 895 905	–		–		–		–		–			–		–		–			
Brick, magnesite fire	2	897	–		–		–		–		–			–		–		–			
Brick, magnezit	2	899 902	–		–		–		–		–			–		–		–			
Brick, Marksa	2	899	–		–		–		–		–			–		–		–			
Brick, metallurgical	2	892 893	–		–		–		–		–			–		–		–			
Brick, metallurgical porous	2	893	–		–		–		–		–			–		–		–			
Brick, Mica	2	892	–		–		–		–		–			–		–		–			
Brick, natural, Sil-O-Cel	2	896	–		–		–		–		–			–		–		–			
Brick, Ordzhonikidze	2	899	–		–		–		–		–			–		–		–			
Brick, Penn. fire	2	905	–		–		–		–		–			–		–		–			
Brick, porous	2	894	–		–		–		–		–			–		–		–			
Brick, porous concrete	2	894	–		–		–		–		–			–		–		–			
Brick, red	2	405 492 898	–		–		–		–		–			–		–		–			
Brick, red hard burned	2	896	–		–		–		–		–			–		–		–			
Brick, red soft burned	2	896	–		–		–		–		–			–		–		–			
Brick, red shamotte	2	405	–		–		–		–		–			–		–		–			
Brick, refractory insulating	2	892	–		–		–		–		–			–		–		–			
Brick, refractory insulating chamotte	2	892	–		–		–		–		–			–		–		–			
Brick, refrax, silicon carbide	2	586 906	–		–		–		–		–			–		–		–			

Substance Name	Thermal Conduc-tivity		Specif. Heat		Thermal Radiative Properties										Thermal Diffu-sivity		Visco-sity		Thermal Expan-sion	
					Emis-sivity		Reflec-tivity		Absorp-tivity		Trans-missiv.									
	V.	Page	V.	Page	V.	Page	V.	Page	V.	Page	V.	Page	V.	Page	V.	Page	V.	Page	V.	Page
Brick, shamotte	2	894 898	–		–		–		–		–		–		–		–		–	
Brick, silica	2	408 489 492 502 894 896 897 898 900 902 904 906	–		–		–		–		–		–		–		–		–	
Brick, silica fire	2	894 895 905	–		–		–		–		–		–		–		–		–	
Brick, silica refractory	2	185	–		–		–		–		–		–		–		–		–	
Brick, silicon carbide	2	895	–		–		–		–		–		–		–		–		–	
Brick, silicous	2	492 902	–		–		–		–		–		–		–		–		–	
Brick, sillimanite	2	902	–		–		–		–		–		–		–		–		–	
Brick, Sil-O-Cel	2	896	–		–		–		–		–		–		–		–		–	
Brick, slag	2	898	–		–		–		–		–		–		–		–		–	
Brick, special, Sil-O-Cel	2	896	–		–		–		–		–		–		–		–		–	
Brick, Star-brand	2	185	–		–		–		–		–		–		–		–		–	
Brick, stillimanite refractory	2	902 903	–		–		–		–		–		–		–		–		–	
Brick, super, Sil-O-Cel	2	896	–		–		–		–		–		–		–		–		–	
Brick, tripolite	2	894	–		–		–		–		–		–		–		–		–	
Brick, vermiculite	2	894	–		–		–		–		–		–		–		–		–	
Brick, white shamotte	2	405	–		–		–		–		–		–		–		–		–	
Brick, zirconia	2	535 895 905	–		–		–		–		–		–		–		–		–	
Brimstone	2	89	5	21	–		–		–		–		–		–		–		–	
Bromic ether		–	6s	4	–		–		–		–		–		–		–		–	
Bromine	3	13	6	7	–		–		–		–		–		11	9	–	
Bromine, monatomic		–	6s	3	–		–		–		–		–		–		–		–	
Bromine chloride		–	6s	3	–		–		–		–		–		–		–		–	
Bromine fluoride		–	6s	4	–		–		–		–		–		–		–		–	
Bromine iodiode		–	6s	54	–		–		–		–		–		–		–		–	
Bromine pentafluoride		–	6s	4	–		–		–		–		–		–		–		–	
Bromobenzene		–	6s	4	–		–		–		–		–		–		–		–	
Bromobenzol		–	6s	4	–		–		–		–		–		–		–		–	
1-Bromobutane		–	6s	4	–		–		–		–		–		–		–		–	
Bromodichloromethane		–	6s	4	–		–		–		–		–		–		–		–	
Bromoethane		–	6s	4	–		–		–		–		–		–		–		–	
Bromoform		–	6s	5	–		–		–		–		–		–		–		–	
Bromomethane		–	6s	5	–		–		–		–		–		–		–		–	
1-Bromo-3-methylbutane		–	6s	5	–		–		–		–		–		–		–		–	

| Substance Name | Thermal Conduc- tivity | | Specif. Heat | | Thermal Radiative Properties | | | | | | | | Thermal Diffu- sivity | | Visco- sity | | Thermal Expan- sion | |
| | | | | | Emis- sivity | | Reflec- tivity | | Absorp- tivity | | Trans- missiv. | | | | | | | |
	V.	Page	V.	Page	V.	Page	V.	Page	V.	Page	V.	Page	V.	Page	V.	Page	V.	Page
1-Bromopropane		–	6s	5		–		–		–		–		–		–		–
Bromotrichloromethane		–	6s	5		–		–		–		–		–		–		–
Bromotrifluoromethane		–		–		–		–		–		–		–	11	104		–
Bromyride	2	569		–		–		–		–		–		–		–		–
Bronze	1	585 586 976 980		–	7	1163	7	918 1167	7	1170		–	10	237		–	12	788 1110 1112
Bronze, aluminum	1	531 532 953		–		–		–		–		–		–		–	12	1091
Bronze, beryllium	1	539		–		–		–		–		–		–		–	12	679
Bronze, manganese		–		–		–		–		–		–		–		–	12	1116
Bronze, phosphor	1	585 586 976		–		–	7	1175		–		–		–		–	12	791 1112
Bronze, silicon	1	973		–		–		–		–		–		–		–		–
Bronze, silver	1	579 980		–		–		–		–		–		–		–		–
Bronze, Navy M	1	977		–		–		–		–		–		–		–		–
Brucite		–		–		–		–		–	8	1662		–		–		–
Burch, photometric sphere white no. 2210		–		–		–	9	490		–		–		–		–		–
Butadiene		–	6s	5		–		–		–		–		–		–		–
1,3-Butadiene		–	6s	5		–		–		–		–		–		–		–
i-Butane	3	139	6	129		–		–		–		–		–	11	109		–
n-Butane	3	141	6	136		–		–		–		–		–	11	114		–
n-Butane-helium, mixture	3	320		–		–		–		–		–		–		–		–
n-Butane-methane, mixture		–		–		–		–		–		–		–	11	357		–
i-Butane-*n*-butane-ethane-helium- methane-nitrogen, mixture		–		–		–		–		–		–		–	11	607		–
1-Butanol		–	6s	6		–		–		–		–		–		–		–
2-Butanol		–	6s	7		–		–		–		–		–		–		–
2-Butanone		–	6s	7		–		–		–		–		–		–		–
2-Butanone, Mek		–	6s	7		–		–		–		–		–		–		–
3-Butanone		–	6s	7		–		–		–		–		–		–		–
1-Butene		–	6s	8		–		–		–		–		–		–		–
2-Butene		–	6s	9		–		–		–		–		–		–		–
cis-2-Butene		–	6s	9		–		–		–		–		–		–		–
trans-2-Butene		–	6s	10		–		–		–		–		–		–		–
1-Butine		–	6s	11		–		–		–		–		–		–		–
2-Butine		–	6s	12		–		–		–		–		–		–		–
1-Butoxybutane		–	6s	11		–		–		–		–		–		–		–
Butyl acetate		–	6s	10		–		–		–		–		–		–		–
Butyl alcohol		–	6s	6		–		–		–		–		–		–		–
sec-Butyl alcohol		–	6s	7		–		–		–		–		–		–		–
tert-Butyl alcohol		–	6s	67		–		–		–		–		–		–		–
Butyl benzene		–	6s	11		–		–		–		–		–		–		–

Substance Name	Thermal Conductivity		Specif. Heat		Thermal Radiative Properties								Thermal Diffusivity		Viscosity		Thermal Expansion	
					Emissivity		Reflectivity		Absorptivity		Transmissiv.							
	V.	Page	V.	Page	V.	Page	V.	Page	V.	Page	V.	Page	V.	Page	V.	Page	V.	Page
tert-Butyl benzene		–	6s	11		–		–		–		–		–		–		–
Butyl bromide		–	6s	4		–		–		–		–		–		–		–
Butyl carbinol		–	6s	72		–		–		–		–		–		–		–
α-Butylene		–	6s	8		–		–		–		–		–		–		–
β-Buylene		–	6s	9		–		–		–		–		–		–		–
cis-β-Butylene		–	6s	9		–		–		–		–		–		–		–
cis-2-Butylene		–	6s	9		–		–		–		–		–		–		–
γ-Butylene		–	6s	67		–		–		–		–		–		–		–
trans-β-Butylene		–	6s	10		–		–		–		–		–		–		–
trans-2-Butylene		–	6s	10		–		–		–		–		–		–		–
2-Butylene		–	6s	9		–		–		–		–		–		–		–
Butyl ethanoate		–	6s	10		–		–		–		–		–		–		–
Butyl ether		–	6s	11		–		–		–		–		–		–		–
Butylethylmethylmethane		–	6s	64		–		–		–		–		–		–		–
sec-Butylethylmethylmethane		–	6s	33		–		–		–		–		–		–		–
1-Butyne		–	6s	11		–		–		–		–		–		–		–
2-Butyne		–	6s	12		–		–		–		–		–		–		–
2-Butyne, crotonylene		–	6s	12		–		–		–		–		–		–		–
Butyric alcohol		–	6s	6		–		–		–		–		–		–		–
Butyric ether		–	6s	37		–		–		–		–		–		–		–
Cabbage		–		–		–		–		–		–	10	630		–		–
Cadmium, Cd	1	45	4	29	7	91	7	93	7	96 98		–	10	17		–	12	40
Cadmium alloys:																		
Cd + Bi	1	517		–		–		–		–		–		–		–		–
Cd + Mg		–	4	294		–		–		–		–		–		–	12	693
Cd + Pb		–		–		–		–		–		–		–		–	12	689
Cd + Sb	1	514		–		–		–		–		–		–		–		–
Cd + Sn	1	521		–		–		–		–		–		–		–		–
Cd + Tl	1	520		–		–		–		–		–		–		–		–
Cd + Zn	1	524		–		–		–		–		–		–		–		–
Cd + Bi + ΣXi	1	941		–		–		–		–		–		–		–		–
Cadmium antimonide, CdSb	1	1264		–		–	8	1282		–		–		–		–		–
Cadmium antimonide + zinc antimonide, mixture	1	1397		–		–		–		–		–		–		–		–
Cadmium-antimony intermetallic compound, CdSb	1	1264		–		–		–		–		–		–		–		–
Cadmium arsenide, Cd₃As₂		–		–		–		–		–		–		–		–	13	744
Cadmium arsenide-zinc arsenide, mixture	1	1396		–		–		–		–		–		–		–		–
Cadmium bromide, CdBr₂		–	5	759		–		–		–		–		–		–		–
Cadmium chloride, CdCl₂		–	5	791		–		–		–		–		–		–		–
Cadmium fluoride, CdF₂		–		–		–		–		–	8	919		–		–	13	1076
Cadmium germanium phosphide, CdGeP₂	2	758		–		–		–		–		–		–		–		–
Cadmium-gold intermetallic compound, CdAu		–		–		–		–		–		–		–		–	12	483

Substance Name	Thermal Conductivity		Specif. Heat		Thermal Radiative Properties								Thermal Diffusivity		Viscosity		Thermal Expansion	
					Emissivity		Reflectivity		Absorptivity		Transmissiv.							
	V.	Page	V.	Page	V.	Page	V.	Page	V.	Page	V.	Page	V.	Page	V.	Page	V.	Page
Cadmium iodide, CdI_2		–	5	491		–		–		–		–		–		–	13	1122
Cadmium–lithium intermetallic compound, CdLi		–		–		–		–		–		–		–		–	12	487 489 490
Cadmium–magnesium intermetallic compound, Cd_3Mg		–		–		–		–		–		–		–		–	12	489 491
Cadmium oxide, CdO		–	5	54	8	216		–		–		–		–		–	13	205
Cadmium selenide, CdSe		–		–		–	8	1108		–	8	1110		–		–	13	1185
Cadmium sulfide, CdS		–	5	650	8	1181 1183	8	1188		–	8	1194		–		–	12	1221
Cadmium telluride, CdTe	1	2167	5	720	8	1239	8	1241		–	8	1244		–		–	13	1243
Cadmium telluride + mercury telluride, mixture	1	1408		–		–		–		–		–		–		–		–
Cadmium tin arsenide, $CdSnAs_2$		–		–		–		–		–		–		–		–	13	752
Cadmium zirconium oxide, $CdO \cdot ZrO_2$		–		–		–		–		–		–		–		–	13	602
Calcia	2	141	5	57		–		‥		–		–		–		–		–
Calcite	2	761		–		–	8	584 1653		–	8	586		–		–		–
Calcium, Ca		–	4	32		–		–		–		–	10	20		–	12	49
Calcium aluminum oxides:																		
$\quad CaO \cdot Al_2O_3$		–	5	1332		–		–		–	8	573		–		–	13	464
$\quad CaO \cdot 2Al_2O_3$		–	5	1335		–		–		–		–		–		–		–
$\quad CaO \cdot 6Al_2O_3$		–		–		–		–		–		–		–		–	13	468
$\quad 3CaO \cdot Al_2O_3$		–	5	1338		–		–		–		–		–		–	13	469
$\quad 3CaO \cdot 5Al_2O_3$		–		–		–		–		–		–		–		–	13	473
$\quad 5CaO \cdot 3Al_2O_3$		–		–		–		–		–		–		–		–	13	474
$\quad 12CaO \cdot 7Al_2O_3$		–	5	1341		–		–		–		–		–		–		–
Calcium aluminum iron oxide, $4CaO \cdot Al_2O_3 \cdot Fe_2O_3$		–		–		–		–		–		–		–		–	13	509
Calcium aluminum silicates:																		
$\quad CaAl_2Si_2O_8$		–	5	1404		–		–		–		–		–		–	13	707
$\quad CaAl_2Si_2O_8 \cdot 2H_2O$		–	5	1407		–		–		–		–		–		–		–
$\quad Ca_2Al_2SiO_7$		–	5	1401		–		–		–		–		–		–	13	727
$\quad Ca_2Al_4Si_8O_{24} \cdot 7H_2O$		–	5	1410		–		–		–		–		–		–		–
Calcium borates:																		
$\quad CaB_2O_4$		–	5	1344		–		–		–		–		–		–		–
$\quad CaB_4O_7$		–	5	1347		–		–		–		–		–		–		–
$\quad Ca_2B_2O_5$		–	5	1350		–		–		–		–		–		–		–
$\quad Ca_3B_2O_6$		–	5	1353		–		–		–		–		–		–		–
Calcium boride, CaB_6		–		–	8	732		–		–		–		–		–		–
Calcium carbide, CaC_2		–	5	405		–		–		–		–		–		–		–
Calcium carbonate, $CaCO_3$	2	759	5	1112		–		–		–		–		–		–	13	637
Calcium chloride, $CaCl_2$		–	5	794		–		–		–		–		–		–		–
Calcium feldspar		–		–		–		–		–		–		–		–	13	707
Calcium fluoride, CaF_2	2	630	5	924	8	921	8	924	8	929	8	931		–		–	13	1025

Substance Name	Thermal Conductivity		Specif. Heat		Thermal Radiative Properties								Thermal Diffusivity		Viscosity		Thermal Expansion	
					Emissivity		Reflectivity		Absorptivity		Transmissiv.							
	V.	Page	V.	Page	V.	Page	V.	Page	V.	Page	V.	Page	V.	Page	V.	Page	V.	Page
Calcium hafnium oxide, $CaO \cdot HfO_2$		–		–		–		–		–		–		–		–	13	501
Calcium iron oxides:																		
$CaC \cdot Fe_2O_3$		–	5	1356		–		–		–		–		–		–	13	503
$2CaO \cdot Fe_2O_3$		–	5	1359		–		–		–		–		–		–	13	506
Calcium-lead intermetallic compounds	1	1271		–		–		–		–		–		–		–		–
Calcium magnesium carbonate, $CaMg(CO_3)_2$		–	5	1115		–		–		–		–		–		–		–
Calcium-magnesium intermetallic compound, $CaMg_2$		–		–		–		–		–		–		–		–	12	493
Calcium magnesium silicates:																		
$CaMgSiO_4$		–		–		–		–		–		–		–		–	13	708
$CaMgSi_2O_6$		–	5	1413		–		–		–		–		–		–	13	708
$Ca_2MgSi_2O_7$		–	5	1416		–		–		–		–		–		–	13	708
$Ca_2Mg_5Si_8O_{23} \cdot H_2O$		–	5	1422		–		–		–		–		–		–		–
$Ca_3MgSi_2O_8$		–	5	1419		–		–		–		–		–		–	13	708
Calcium magnesium silicate, merwinite		–		–		–		–		–		–		–		–	13	708
Calcium magnesium silicate, monticellite		–		–		–		–		–		–		–		–	13	708
Calcium magnesium tungsten oxide, $2CaO \cdot MgO \cdot WO_3$		–		–		–		–		–		–		–		–	13	584
Calcium molybdenum oxide, $CaO \cdot MoO_3$		–	5	1362		–		–		–		–		–		–	13	517
Calcium oxide, CaO	2	141	5	57		–	8	218		–	8	220		–		–	13	208
Calcium oxide + magnesium oxide + ΣXi, mixture	2	477		–		–		–		–		–		–		–		–
Calcium oxide + silicon oxide, mixture	2	407		–		–		–		–		–		–		–		–
Calcium oxide + silicon oxide + ΣXi, mixture	2	501		–		–		–		–		–		–		–		–
Calcium oxide + uranium dioxide, mixture	2	426		–		–		–		–		–	10	438		–		–
Calcium oxide + zirconium oxide, mixture	2	442		–		–		–		–		–	10	450		–		–
Calcium oxide + zirconium oxide + ΣXi, mixture	2	531		–		–		–		–		–		–		–		–
Calcium phosphate + lithium carbonate + magnesium carbonate,	2	763		–		–		–		–		–		–		–		–
Calcium silicates:																		
$CaSiO_3$		–	5	1365		–		–		–		–		–		–	13	705
Ca_2SiO_4		–	5	1368		–	8	618		–		–		–		–	13	705
Ca_3SiO_5		–	5	1371		–		–		–		–		–		–	13	705
$Ca_3Si_2O_7$		–	5	1374		–		–		–		–		–		–	13	705
Calcium stannate, $CaSnO_3$	2	264		–		–		–		–		–		–		–		–
Calcium stannide, Ca_2Sn	1	1273		–		–		–		–		–		–		–		–
Calcium sulfates:																		
$CaSO_4$		–	5	1182		–	8	627		–	8	629		–		–		–
$CaSO_4 \cdot 1/2H_2O$		–	5	1185		–		–		–		–		–		–		–
$CaSO_4 \cdot 2H_2O$		–	5	1188		–		–		–		–		–		–		–
Calcium sulfide, CaS		–	5	653		–		–		–		–		–		–	13	1239
Calcium-tin intermetallic compound, Ca_2Sn	1	1273		–		–		–		–		–		–		–		–

Substance Name	Thermal Conductivity		Specif. Heat		Thermal Radiative Properties								Thermal Diffusivity		Viscosity		Thermal Expansion	
					Emissivity		Reflectivity		Absorptivity		Transmissiv.							
	V.	Page	V.	Page	V.	Page	V.	Page	V.	Page	V.	Page	V.	Page	V.	Page	V.	Page
Calcium titanium oxides:																		
CaO·TiO₂	2	267	5	1377	–		–		–		–		–			–		–
3CaO·2TiO₂		–	5	1380	–		–		–		–		–			–		–
Calcium tungsten oxide, CaO·WO₃	2	270	5	1383	–		–			–	8	663	10	415		–	13	580
Calcium uranium oxide, CaO·UO₃		–	5	1386	–		–		–		–		–			–		–
Calcium vanadium oxides:																		
CaC·V₂O₅		–	5	1389	–		–		–		–		–			–		–
2CaO·V₂O₅		–	5	1392	–		–		–		–		–			–		–
3CaO·V₂O₅		–	5	1395	–		–		–		–		–			–		–
Calcium zirconium oxide, CaO·ZrO₂		–	5	1398	–		–		–		–		–			–	13	603
Calcium zirconium silicate, CaZrSiO₅		–		–	–		–		8	616	–		–			–		–
Carbides, miscellaneous		–		–	8	847 849 851	8	854	–		–		–			–		–
Carbomethene		–	6s	57	–		–		–		–		–			–		–
Carbon	2	5		–	8	5 8 10 12 14 16	8	18 20 22 24 25	–		8	27	10	21		–		–
Carbon, atomic		–	6s	12	–		–		–		–		–			–		–
carbon, glassy C		–		–	–		–		–		–		–			–	13	16
Carbon, graphite		–	5	9	8	30 31 38 40 42 44 51 57	8	59 61 63 65 70 71	8	74 76	–		–			–		–
Carbon, graphitized		–		–	–		–		–		–		–			–	13	130
Carbon black, channel	2	764		–	–		–		–		–		–			–		–
Carbon black, graphitized	2	60		–	–		–		–		–		–			–		–
Carbon, diamond		–	5	4	–		–		–		–		–			–		–
Carbon dichloride		–	6s	90	–		–		–		–		–			–		–
Carbon dioxide, CO₂	3	145	6	143	–		–		–		–		–		11	119		–
Carbon dioxide-carbon monoxide-hydrogen-methane-nitrogen, mixture		–		–	–		–		–		–		–		11	620		–
Carbon dioxide-carbon monoxide-hydrogen-methane- nitrogen-oxygen, mixture		–		–	–		–		–		–		–		11	621		–
Carbon dioxide-carbon monoxide-hydrogen-methane- nitrogen-oxygen-heavier hydrocarbons, mixture		–		–	–		–		–		–		–		11	622		–
Carbon dioxide-carbon monoxide-hydrogen-nitrogen-oxygen, mixture		–		–	–		–		–		–		–		11	623		–
Carbon dioxide-ethylene, mixture	3	389		–	–		–		–		–		–			–		–
Carbon dioxide-helium, mixture	3	322		–	–		–		–		–		–		11	297		–
Carbon dioxide-hydrogen, mixture	3	391		–	–		–		–		–		–		11	366		–
Carbon dioxide-hydrogen chloride, mixture		–		–	–		–		–		–		–		11	501		–

Substance Name	Thermal Conductivity		Specif. Heat		Thermal Radiative Properties								Thermal Diffusivity		Viscosity		Thermal Expansion	
					Emissivity		Reflectivity		Absorptivity		Transmissiv.							
	V.	Page	V.	Page	V.	Page	V.	Page	V.	Page	V.	Page	V.	Page	V.	Page	V.	Page
Carbon dioxide-hydrogen-nitrogen-oxygen, mixture		–		–	–		–		–		–			–	11	595		–
Carbon dioxide-hydrogen-oxygen, mixture		–		–	–		–		–		–			–	11	584		–
Carbon dioxide-krypton, mixture		–		–	–		–		–		–			–	11	331		–
Carbon dioxide-neon, mixture	3	385		–	–		–		–		–			–	11	334		–
Carbon dioxide-methane, mixture		–		–	–		–		–		–			–	11	369		–
Carbon dioxide-nitrogen, mixture	3	396		–	–		–		–		–			–	11	376		–
Carbon dioxide-nitrogen-oxygen, mixture	3	497		–	–		–		–		–			–	11	585		–
Carbon dioxide-nitrous oxide, mixture		–		–	–		–		–		–			–	11	383		–
Carbon dioxide-oxygen, mixture	3	401		–	–		–		–		–			–	11	385		–
Carbon dioxide-propane, mixture	3	403		–	–		–		–		–			–	11	387		–
Carbon dioxide-steam, mixture	3	466		–	–		–		–		–			–		–		–
Carbon dioxide-sulfur dioxide, mixture		–		–	–		–		–		–			–	11	503		–
Carbon fluoride, CF_6		–	6s	44	–		–		–		–			–		–		–
Carbon monoxide, CO	3	151	6s / 6	15 / 152	–		–		–		–			–	11	125		–
Carbon monoxide-ethylene, mixture		–		–	–		–		–		–			–	11	389		–
Carbon monoxide-hydrogen, mixture	3	405		–	–		–		–		–			–	11	391		–
Carbon monoxide-nitrogen, mixture		–		–	–		–		–		–			–	11	393		–
Carbon monoxide-oxygen, mixture		–		–	–		–		–		–			–	11	397		–
Carbon nitride + uranium cerium, mixture		–		–	–		–		–		–			–		–	13	1163
Carbon oxychloride		–	6s	74	–		–		–		–			–		–		–
Carbon oxyfluoride		–	6s	16	–		–		–		–			–		–		–
Carbon-oxygen, mixture	2	764		–	–		–		–		–			–		–		–
Carbon oxysulfide		–	6s	16	–		–		–		–			–		–		–
Carbon resistor graphite	2	73		–	–		–		–		–			–		–		–
Carbon + silicon carbide, mixture		–	5	1276	–		–		–		–			–		–		–
Carbon + silicon carbide + ΣXi, mixture		–	5	1279	–		–		–		–			–		–		–
Carbon + silicon carbide + zirconium boride, mixture		–		–	–		–		–		–		10	538		–		–
Carbon sulfides:																		
CS		–	6s	15	–		–		–		–			–		–		–
CS_2		–	6s	13	–		–		–		–			–		–	13	1239
Carbon tetrabromide		–	6s	15	–		–		–		–			–		–		–
Carbon tetrachloride	3	156	6	159	–		–		–		–			–	11	129		–
Carbon tetrafluoride		–		–	–		–		–		–			–	11	131		–
Carbon tetrachloride-dichloromethane, mixture		–		–	–		–		–		–			–	11	506		–
Carbon tetrachloride-methane, mixture		–		–	–		–		–		–			–	11	401		–
Carbon tetrachloride-methanol, mixture		–		–	–		–		–		–			–	11	510		–
Carbon tetrachloride-octamethylcyclo-tetrasiloxane, mixture		–		–	–		–		–		–			–	11	399		–

| Substance Name | Thermal Conductivity | | Specif. Heat | | Thermal Radiative Properties | | | | | | | | | | Thermal Diffusivity | | Viscosity | | Thermal Expansion | |
| | | | | | Emissivity | | Reflectivity | | Absorptivity | | Transmissiv. | | | | | | | | | |
	V.	Page	V.	Page	V.	Page	V.	Page	V.	Page	V.	Page	V.	Page	V.	Page	V.	Page	V.	Page
Carbon tetrachloride-isopropyl alcohol, mixture		–		–		–		–		–		–		–			11	508		–
Carbon tetrachloride-sulfur hexafluoride, mixture		–		–		–		–		–		–		–			11	406		–
Carbon tetradeuteride		–	6s	58		–		–		–		–		–				–		–
Carbon + titanium carbide, mixture		–		–		–		–		–		–		–				–	13	953
Carbon + uranium carbide		–		–		–		–		–		–		–				–	13	954
Carbon + volatile materials	2	765		–		–		–		–		–		–				–		–
Carbon + zirconium boride, mixture		–				–		–		–		–	10	534				–		–
Carbon + zirconium carbide, mixture		–		–		–		–		–		–		–				–	13	957
Carbonates, miscellaneous		–		–		–	8	592		–		–		–				–		–
Carbonyl chloride		–	6s	74		–		–		–		–		–				–		–
Carbonyl chloride fluoride		–	6s	15		–		–		–		–		–				–		–
Carbonyl fluoride		–	6s	16		–		–		–		–		–				–		–
Carbonyl sulfide		–	6s	16		–		–		–		–		–				–		–
Carborundum	2	553 555 596		–		–		–		–		–		–				–		–
Carboxybenzene		–	6s	2		–		–		–		–		–				–		–
Cardboard	2	1109		–		–		–		–		–		–				–		–
Carnegiete		–		–		–		–		–		–		–				–	13	728
Carrot		–		–		–		–		–		–	10	631				–		–
Cassiopeium	1	198	4	121		–		–		–		–		–				–		–
Cassiterite		–		–		–	8	451		–		–		–				–	13	388
Castor oil		–		–		–		–		–		–	10	632				–		–
Cedar	2	1062		–		–		–		–		–		–				–		–
Ceiba, kapok	2	1077		–		–		–		–		–		–				–		–
Celestite		–		–		–		–		–	8	1701		–				–		–
Celkate T-21		–		–		–		–		–		–	10	440				–		–
Cellulose acetate		–		–		–		–		–		–		–				–	13	1407
Cellulose acetate, Tenite I		–		–		–		–		–		–		–				–	13	1407
Cellulose acetate, Tenite II		–		–		–		–		–		–		–				–	13	1407
Cellulose proprionate		–		–		–		–		–		–		–				–	13	1408
Celtium	1	138	4	87		–		–		–		–		–				–		–
Cement, Portland	2	861		–		–		–		–		–		–				–		–
Cement, Portland + slag, mixture	2	861		–		–		–		–		–		–				–		–
Ceramics:																				
Al₂O₃ + ΣXi		–		–		–		–		–		–		–				–	13	1278
BaO·TiO₂		–		–		–		–		–		–		–				–	13	1280
Ceramag		–		–		–		–		–		–		–				–	13	1283
Ceramag 7A		–		–		–		–		–		–		–				–	13	1284
Ceramag 9		–		–		–		–		–		–		–				–	13	1284
Ceramag 23B		–		–		–		–		–		–		–				–	13	1284
Ceramag 27		–		–		–		–		–		–		–				–	13	1284

| Substance Name | Thermal Conductivity | | Specif. Heat | | Thermal Radiative Properties | | | | | | | | | | | Thermal Diffusivity | | Viscosity | | Thermal Expansion | |
| | | | | | Emissivity | | Reflectivity | | Absorptivity | | Transmissiv. | | | | | | | | | | | |
	V.	Page	V.	Page	V.	Page	V.	Page	V.	Page	V.	Page	V.	Page	V.	Page	V.	Page
Ceramics: (continued)																		
Ceramag 2817		–		–		–		–		–		–		–		–	13	1284
Ceramag 7441		–		–		–		–		–		–		–		–	13	1284
Miscellaneous	2	915		–		–		–		–		–		–		–	13	1291
Cerium, Ce	1	50	4	36		–		–		–		–	10	43		–	12	53
Cerium alloys:																		
Ce + Mg				–		–		–		–		–				–	12	702
Ce + Th				–		–		–		–		–				–	12	702
Cerium arsenide, CeAs		–		–		–		–		–		–				–	13	752
Cerium boride, CeB$_6$		–		–	8	722		–		–		–		–		–		–
Cerium carbides:																		
CeC$_2$		–		–		–		–		–		–		–		–	13	935
Ce$_2$C$_3$		–		–		–		–		–		–		–		–	13	935
Cerium oxides:																		
CeO$_2$	2	144	5	60	8	225	8	227		–		–		–		–	13	212
Ce$_2$O$_3$		–	5	64		–		–		–		–		–		–		–
Cerium sulfides:																		
CeS		–	5	656		–		–		–		–		–		–	13	1239
Ce$_2$S$_3$		–	5	659	8	1231		–		–		–		–		–	13	1239
Cerium trifluoride, CeF$_3$		–	5	927		–		–		–		–		–		–		–
Cerium-indium intermetallic compound, CeIn$_3$		–		–		–		–		–		–		–		–	12	496 498 499
Cerium dioxide + magnesium oxide, mixture	2	350		–		–		–		–		–		–		–		–
Cerium dioxide + uranium dioxide, mixture	2	353		–		–		–		–		–		–		–		–
Cerium-palladium intermetallic compound, CePd$_3$		–		–		–		–		–		–		–		–	12	497 498 500
Cerium-ruthenium intermetallic compound, CeRu$_2$		–		–		–		–		–		–		–		–	12	501
Cerium-tin intermetallic compound, CeSn$_3$		–		–		–		–		–		–		–		–	12	504
Cermets:																		
Al$_2$O$_3$ + Al		–		–		–		–		–		–		–		–	13	1306
Al$_2$O$_3$ + Cr	2 1	707 1419		–		–		–		–		–		–		–		–
Al$_2$O$_3$ + Mo		–		–		–		–		–		–	10	566		–		–
Al$_2$O$_3$ + NiAl		–		–	8	1358 1359	8	1363		–		–		–		–		–
B$_4$C + Al	2	717		–		–		–		–		–		–		–		–
BeO + Be	2 1	708 1416	5	1243 1246		–		–		–		–		–		–	13	1313
BeO + Be, QMV		–	5	1243		–		–		–		–		–		–		–
BeO + Be, YB 9052		–	5	1243		–		–		–		–		–		–		–
BeO + Be, YB 9054		–	5	1243		–		–		–		–		–		–		–

Substance Name	Thermal Conductivity		Specif. Heat		Thermal Radiative Properties								Thermal Diffusivity		Viscosity		Thermal Expansion	
					Emissivity		Reflectivity		Absorptivity		Transmissiv.							
	V.	Page	V.	Page	V.	Page	V.	Page	V.	Page	V.	Page	V.	Page	V.	Page	V.	Page
Cermets:																		
(continued)																		
BeO + BeTa		–		–	8	1377 1378	8	1382		–		–		–		–		–
BeO + Mo		–	5	1252		–		–		–		–		–		–		–
BeO + MoBe$_{12}$		–	5	1255		–		–		–		–		–		–		–
BeO + NbBe$_{12}$		–	5	1258		–		–		–		–		–		–		–
BeO + TaBe$_{12}$		–	5	1261		–		–		–		–		–		–		–
BeO + TiBe$_{12}$		–	5	1264		–		–		–		–		–		–		–
BeO + ZrBe$_{13}$		–	5	1267		–		–		–		–		–		–		–
CrO$_3$ + TiCr		–		–	8	1385 1386	8	1390		–		–		–		–		–
FeSi$_2$ + FeAl$_2$		–		–		–		–		–		–	10	529		–		–
Gd$_2$O$_3$ + Tb		–		–		–		–		–		–		–		–	13	1341
Haynes LT-1		–		–	8	1356		–		–		–		–		–		–
Haynes LT-1B		–		–	8	1356		–		–		–		–		–		–
Haynes LT-2		–		–	8	1375		–		–		–		–		–		–
HfO$_2$ + Fe		–		–		–		–		–		–		–		–	13	1317
MoO$_2$ + Ni		–		–		–	8	1425		–		–		–		–		–
MoO$_3$ + Ni		–		–		–	8	1425		–		–		–		–		–
Na$_2$O + Na	2 1	721 1432		–		–		–		–		–		–		–		–
SiO$_2$ + Al		–		–		–	8	1428		–		–		–		–		–
SiO$_2$ + Cr		–		–		–		–		–	8	1401		–		–		–
SrTiO$_3$ + Co	2	722		–		–		–		–		–		–		–		–
Ta$_2$O$_5$ + Be$_{12}$Ta		–		–	8	1403 1404	8	1406		–		–		–		–		–
ThO$_2$ + Mo	1	1429		–		–		–		–		–		–		–		–
ThO$_2$ + Ni		–		–	8	1408		–		–		–		–		–		–
ThO$_2$ + W	1	1439		–		–		–		–		–		–		–		–
TiB$_2$ + Ni		–		–		–	8	1435		–		–		–		–		–
TiB$_2$ + Ti		–		–	8	1410	8	1437		–		–		–		–		–
TiC + Co	2	725		–				–		–		–		–		–	13	1341
TiC + steel T-420-G		–		–		–		–		–		–		–		–	13	1341
TiC + steel T-520-G		–		–		–		–		–		–		–		–	13	1341
TiO$_2$ + Ti		–		–		–	8	1439		–		–		–		–		–
Ti$_2$O$_3$ + Ti		–		–		–		–		–		–		–		–	13	1341
TiO$_2$ + TiCr$_2$		–		–	8	1419 1420	8	1421		–		–		–		–		–
WC + Co		–		–				–		–		–		–		–	13	1322
UC + U	2	731		–		–		–		–		–		–		–	13	1342
UO$_2$ + Cr	2	732		–		–		–		–		–		–		–	13	1342
UO$_2$ + Mo	2	735		–		–		–		–		–		–		–	13	1326
UO$_2$ + Nb	2	738		–		–		–		–		–		–		–		–
UO$_2$ + stainless steel	2	741		–		–		–		–		–		–		–	13	1330

Substance Name	Thermal Conductivity		Specif. Heat		Thermal Radiative Properties								Thermal Diffusivity		Viscosity		Thermal Expansion	
					Emissivity		Reflectivity		Absorptivity		Transmissiv.							
	V.	Page	V.	Page	V.	Page	V.	Page	V.	Page	V.	Page	V.	Page	V.	Page	V.	Page
Cermets: (continued)																		
UO_2 + U	2 1	744 1442	–		–		–		–		–		–		–		–	
UO_2 + Zr	2	746	–		–		–		–		–		–		–		13	1342
USi_3 + W	–		–		–		–		–		–		–		–		13	1342
ZrB_2 + Cr	–		–		–		–		–		–		–		–		13	1342
ZrH_2 + U	–		–		–		–		–		–		10	540	–		–	
ZrO_2 + Al	–		–		–		8	1442	–		–		–		–		–	
ZrO_2 + Ti	2	749	5	1285	–		–		–		–		–		–		13	1333
ZrO_2 + Zr	2 1	752 1444	–		–		–		–		–		–		–		13	1337
Al_2O_3 + Cr + ΣXi	–		–		8	1355	–		–		–		–		–		13	1309
Al_2O_3 + W + ΣXi	–		–		8	1375	–		–		–		–		–		–	
BeO + Be + Mo	2	711	5	1249	–		–		–		–		–		–		13	1341
BeO + Be + Si	2	714	–		–		–		–		–		–		–		13	1341
NiO + NiAl + ΣXi	–		–		8	1393 1394	8	1398	–		–		–		–		–	
TiC + Co + NbC	2	726	–		–		–		–		–		–		–		–	
TiC + Ni + ΣXi	–		–		8	1412 1415	–		–		–		–		–		13	1319
TiC + Ni + NbC	2	730	–		–		–		–		–		–		–		–	
ZrO_2 + Y_2O_3 + Zr	2	753	–		–		–		–		–		–		–		–	
TiC + Ni + Mo + NbC	2	727	–		–		–		–		–		–		–		–	
Cesium, Cs	1	54	4	40	–		–		–		–		10	44	–		12	60
Cesium aluminum silicate, $CsAlSi_2O_6$	–		–		–		–		–		–		–		–		13	709
Cesium aluminum sulfate dodecahydrate, $CsAl(SO_4)_2 \cdot 12$	–		5	1191	–		–		–		–		–		–		–	
Cesium bromide, CsBr	2	565	–		–		8	737	–		8	739	–		–		13	801
Cesium chloride, CsCl	–		5	797	–		–		–		–		–		–		13	973
Cesium chloride + ΣXi, mixture	–		–		–		–		–		–		–		–		13	1015
Cesium fluoride, CsF	–		–		–		–		–		–		–		–		13	1076
Cesium hydrogen fluoride, $CsHF_2$	–		5	931	–		–		–		–		–		–		–	
Cesium iodide, CsI	2	561	5	494	–		8	995	–		8	997	–		–		13	1098
Cesium sulfate, Cs_2SO_4	–		–		–		–		–		–		–		–		13	730
Cetane	–		6s	43	–		–		–		–		–		–		–	
Charcoal	2	1157	–		–		–		–		–		–		–		–	
Charcoal, powder	2	1040	–		–		–		–		–		–		–		–	
Chinone	–		6s	2	–		–		–		–		–		–		–	
Chlorides, miscellaneous	–		–		–		8	905	–		8	907	–		–		–	
Chlorinated hydrochloric ether	–		6s	27	–		–		–		–		–		–		–	
Chlorine, Cl_2	3	17	6	11	–		–		–		–		–		11	11	–	
Chlorine, monatomic	–		6s	17	–		–		–		–		–		–		–	
Chlorine cyanide	–		6s	24	–		–		–		–		–		–		–	
Chlorine dioxide	–		6s	19	–		–		–		–		–		–		–	

45

Substance Name	Thermal Conductivity		Specif. Heat		Thermal Radiative Properties								Thermal Diffusivity		Viscosity		Thermal Expansion	
					Emissivity		Reflectivity		Absorptivity		Transmissiv.							
	V.	Page	V.	Page	V.	Page	V.	Page	V.	Page	V.	Page	V.	Page	V.	Page	V.	Page
Chlorine fluoride		–	6s	19		–		–		–		–		–		–		–
Chlorine iodide		–	6s	54		–		–		–		–		–		–		–
Chlorine monoxide		–	6s	20		–		–		–		–		–		–		–
Chlorine oxide		–	6s	20		–		–		–		–		–		–		–
Chlorine trifluoride		–	6s	20		–		–		–		–		–		–		–
Chlorobenzene		–	6s	21		–		–		–		–		–		–		–
m-Chlorobenzoic acid		–	6s	21		–		–		–		–		–		–		–
o-Chlorobenzoic acid		–	6s	21		–		–		–		–		–		–		–
p-Chlorobenzoic acid		–	6s	21		–		–		–		–		–		–		–
Chlorodifluoromethane	3	197	6	218		–		–		–		–		–	11	133		–
Chlorodifluoromethane, monodeuterated		–	6s	21		–		–		–		–		–		–		–
Chlorodiphenylmethane		–	6s	21		–		–		–		–		–		–		–
m-Chlorodracylic acid		–	6s	21		–		–		–		–		–		–		–
o-Chlorodracylic acid		–	6s	21		–		–		–		–		–		–		–
p-Chlorodracylic acid		–	6s	21		–		–		–		–		–		–		–
Chloroethane		–	6s	21		–		–		–		–		–		–		–
Chlorofluorocarbonyl		–	6s	15		–		–		–		–		–		–		–
Chlorofluoromethane		–	6s	21		–		–		–		–		–		–		–
Chloroform, $CHCl_3$	3	161	6	166		–		–		–		–		–	11	138		–
Chloroform-ethyl ether, mixture	3	470		–		–		–		–		–		–		–		–
Chloroformyl chloride		–	6s	74		–		–		–		–		–		–		–
Chloromethane	3	227	6	257		–		–		–		–		–		–		–
1-Chloro-3-methylbutane		–	6s	22		–		–		–		–		–		–		–
Chloromethylidyne		–	6s	22		–		–		–		–		–		–		–
1-Chloro-2-methylpropane		–	6s	22		–		–		–		–		–		–		–
Chloropentalfluoroethane		–		–		–		–		–		–		–	11	140		–
1-Chloropropane		–	6s	22		–		–		–		–		–		–		–
Chlorosilane		–	6s	22		–		–		–		–		–		–		–
α-Chlorotoluene		–	6s	22		–		–		–		–		–		–		–
ω-Chlorotoluene		–	6s	22		–		–		–		–		–		–		–
Chlorotribromomethane		–	6s	22		–		–		–		–		–		–		–
Chlorotrifluoromethane	3	191		–		–		–		–		–		–	11	145		–
Chromalloy, W-2	1	1324		–	9	1333 1335 1337 1350		–	9	1345		–		–		–		–
Chromatone		–		–		–	9	494		–		–		–		–		–
Chromium, Cr	1	60	4	44	7	101 103 106	7	110 113	7	115 118	7	120	10	45		–	12	61
Chromium alloys:																		
Cr + Al		–	4	304		–		–		–		–		–		–		–
Cr + Co		–		–		–		–		–		–		–		–	12	705
Cr + Fe		–	4	307		–		–		–		–		–		–	12	708
Cr + Mn		–	4	311		–		–		–		–		–		–	12	714

Substance Name	Thermal Conductivity		Specif. Heat		Thermal Radiative Properties								Thermal Diffusivity		Viscosity		Thermal Expansion	
					Emissivity		Reflectivity		Absorptivity		Transmissiv.							
	V.	Page	V.	Page	V.	Page	V.	Page	V.	Page	V.	Page	V.	Page	V.	Page	V.	Page
Chromium alloys: (continued)																		
Cr + Mo		–		–		–		–		–		–		–		–	12	713
Cr + Ni	1	525		–		–		–		–		–		–		–	12	719
Cr + Ni, Vickers F. D. P.		–		–	7	1221		–		–		–		–		–		–
Cr + Si		–		–		–		–		–		–		–		–	12	721
Cr + Sn		–		–		–		–		–		–		–		–	12	725
Cr + Ti		–		–		–		–		–		–		–		–	12	726
Cr + V		–		–		–		–		–		–		–		–	12	729
Cr + Al + ΣXi		–	4	517		–		–		–		–		–		–		–
Cr + Fe + ΣXi	1	944	4	520		–		–		–		–		–		–	12	1060
Cr + Fe + ΣXi, aluminothermic chromium		–	4	520		–		–		–		–		–		–		–
Cr + Fe + ΣXi, Russian, ferrochromium	1	945	4	520		–		–		–		–		–		–		–
Cr + Si + ΣXi		–		–		–		–		–		–		–		–	12	1064
Chromium + aluminum oxide, cermet	1	1419		–		–		–		–		–		–		–		–
Chromium borides:																		
CrB		–	5	335		–	8	734		–		–		–		–	13	796
CrB_2		–	5	338	8	731		–		–		–		–		–	13	796
Cr_2B		–		–		–	8	734		–		–		–		–		–
Chromium carbides:																		
Cr_3C_2		–	5	408	8	852		–		–		–		–		–	13	845
Cr_4C		–	5	414		–		–		–		–		–		–		–
Cr_5C_2		–	5	411		–		–		–		–		–		–		–
Cr_7C_3		–	5	417	8	852		–		–		–		–		–		–
Chromium chlorides:																		
$CrCl_2$		–	5	800		–		–		–		–		–		–		–
$CrCl_3$		–	5	803		–		–		–		–		–		–		–
Chromium-iron intermetallic compound, CrFe		–		–		–		–		–		–		–		–	12	507
Chromium nitrides:																		
CrN		–		–	8	1087		–		–		–		–		–		–
Cr_2N		–		–	8	1087		–		–		–		–		–		–
Chromium oxides:																		
CrO_3		–		–		–	8	236		–		–		–		–		–
Cr_2O_3		–	5	67	8	231 233	8	236		–		–		–		–	13	217
Chromium oxide + magnesium oxide + ΣXi, mixture	2	473 480		–		–		–		–		–		–		–		–
Chromium oxide + titanium chromium compound, cermet		–		–	8	1385 1386	8	1390		–		–		–		–		–
Chromium oxide + yttrium oxide powders		–		–	8	509 570		–		–		–		–		–		–

Substance Name	Thermal Conductivity		Specif. Heat		Thermal Radiative Properties										Thermal Diffusivity		Viscosity		Thermal Expansion	
					Emissivity		Reflectivity		Absorptivity		Transmissiv.									
	V.	Page	V.	Page	V.	Page	V.	Page	V.	Page	V.	Page	V.	Page	V.	Page	V.	Page
Chromium silicides:																		
CrSi		–	5	565	8	1140 1142		–		–		–		–		–	13	1211
CrSi₂		–	5	568	8	1140 1142		–		–		–		–		–	13	1211
Cr₃Si		–	5	559	8	1139 1140 1142	8	1144		–		–		–		–	13	1195
Cr₃Si₂		–		–	8	1140 1142		–		–		–		–		–	13	1211
Cr₅Si₃		–	5	562		–		–		–		–		–		–		–
Chromium telluride, CrTe		–		–		–		–		–		–		–		–	13	1248
Chromium tungsten oxide, Cr₂O₃·WO₃		–		–		–		–		–		–		–		–	13	586
Chromium vanadium oxide, Cr₂O₃·V₂O₅		–		–		–		–		–		–		–		–	13	595
Cinnamene		–	6s	84		–		–		–		–		–		–		–
Clay				–		–		–		–		–	10	546		–		–
Clay, Ashkhabad	2	804 805		–		–		–		–		–		–		–		–
Clay, Beskhudnikov	2	804		–		–		–		–		–		–		–		–
Clay, chamotte	2	804		–		–		–		–		–		–		–		–
Clay, Dixie		–		–		–		–		–		–	10	546		–		–
Clay, fire	2	804		–		–		–		–		–		–		–		–
Clay, fire aluminous	2	489		–		–		–		–		–		–		–		–
Clay, fire light weight	2	403 404		–		–		–		–		–		–		–		–
Clay, fire pressed	2	403		–		–		–		–		–		–		–		–
Clay, Kuchin	2	804		–		–		–		–		–		–		–		–
Clay + magnesium oxide, mixture	2	374		–		–		–		–		–		–		–		–
Climax	1	1198 1213		–		–		–		–		–		–		–		–
Clinoenstatite		–		–		–		–		–		–		–		–	13	717
Coal, angren brown	2	808		–		–		–		–		–		–		–		–
Coal, brown		–		–		–		–		–		–	10	35		–		–
Coal, donets gas	2	808		–		–		–		–		–		–		–		–
Coal, donets anthracite	2	808		–		–		–		–		–		–		–		–
Coal, gas		–		–		–		–		–		–	10	22 35		–		–
Coal, tar fractions	2	1158		–		–		–		–		–		–		–		–
Coating, acrylic on:																		
Aluminum substrate		–		–	9	1108	9	1110		–		–		–		–		–
Aluminum oxide substrate		–		–	9	1109		–		–		–		–		–		–
Coating, alkyd on aluminum substrate		–		–		–	9	1111		–		–		–		–		–
Coating, aluminum on:																		
Aluminum substrate		–		–	9	580	9	592	9	610		–		–		–		–
Copper substrate		–		–		–	9	594		–		–		–		–		–
Epoxy substrate		–		–		–	9	602	9	610		–		–		–		–
Fabric substrate		–		–	9	586		–		–		–		–		–		–

Substance Name	Thermal Conductivity		Specif. Heat		Thermal Radiative Properties								Thermal Diffusivity		Viscosity		Thermal Expansion	
					Emissivity		Reflectivity		Absorptivity		Transmissiv.							
	V.	Page	V.	Page	V.	Page	V.	Page	V.	Page	V.	Page	V.	Page	V.	Page	V.	Page
Coating, aluminum on: (continued)																		
Glass substrate	–		–		–		9	591 602 607	–		–		–		–		–	
Iron substrate	–		–		9	580	–		–		–		–		–		–	
Lacquer substrate	–		–		–		9	600	–		–		–		–		–	
Mylar substrate	–		–		9	580	9	592	9	609	–		–		–		–	
Polyester substrate	–		–		9	580	–		–		–		–		–		–	
Polyurethane substrate	–		–		–		9	602	–		–		–		–		–	
Quartz substrate	–		–		9	581	9	592 602	–		–		–		–		–	
Silver substrate	–		–		–		9	603	–		–		–		–		–	
Stainless steel substrate	–		–		9	580	9	602	9	609 612	–		–		–		–	
Teflon substrate	–		–		9	586	–		9	610	–		–		–		–	
Coating, aluminum + magnesium on glass substrate	–		–		–		9	613	–		–		–		–		–	
Coating, aluminum oxide on:																		
Aluminum substrate	–		–		9	785	9	794 796 799 800	9	803	–		–		–		–	
Gold substrate	–		–		–		–		9	803	–		–		–		–	
Inconel substrate	–		–		9	788 792	–		–		–		–		–		–	
Mild steel substrate	–		–		9	788 792	–		–		–		–		–		–	
Molybdenum substrate	–		–		9	785 788	9	796	9	803	–		–		–		–	
Nimonic 75 substrate	–		–		9	788	–		–		–		–		–		–	
Niobium substrate	–		–		9	785	–		–		–		–		–		–	
Silicon monoxide substrate	–		–		–		9	1077	–		–		–		–		–	
Silver substrate	–		–		9	785 788	9	796	9	802 803	–		–		–		–	
Stainless steel substrate	–		–		9	788 790 792	9	796	9	804	–		–		–		–	
Unknown substrate	–		–		–		–		–		9	807	–		–		–	
Coating, aluminum oxide + aluminum titanate on Nb-1Zr substrate	–		–		9	808	–		–		–		–		–		–	
Coating, alundum on niobium substrate	–		–		9	785	–		–		–		–		–		–	
Coating, AN-L-29 on:																		
Aluminum substrate	–		–		9	1124	–		9	1127	–		–		–		–	
Dow metal substrate	–		–		9	1124	–		9	1127	–		–		–		–	
Coating, AN-TT-V-116 on:																		
Aluminum substrate	–		–		9	1124	–		9	1127	–		–		–		–	
Dow metal substrate	–		–		9	1124	–		9	1127	–		–		–		–	
Coating, anodized aluminum on stainless steel substrate	–		–		–		–		9	1226 1230 1231	–		–		–		–	

Substance Name	Thermal Conduc- tivity		Specif. Heat		Thermal Radiative Properties								Thermal Diffu- sivity		Visco- sity		Thermal Expan- sion	
					Emis- sivity		Reflec- tivity		Absorp- tivity		Trans- missiv.							
	V.	Page	V.	Page	V.	Page	V.	Page	V.	Page	V.	Page	V.	Page	V.	Page	V.	Page
Coating, antimony on:																		
Aluminum substrate		–		–		–		–		–	9	615		–		–		–
Glass substrate		–		–		–	9	614		–		–		–		–		–
Stilbene substrate		–		–		–		–		–	9	615		–		–		–
Sb + Cu	1	495		–		–		–		–		–	10	227		–		–
Coating, antimony black on:																		
Celluose nitrate substrate		–		–		–		–		–	9	1172		–		–		–
KRS-5 substrate		–		–		–		–		–	9	1172		–		–		–
Coatings, applied, nonmetallic	2	1009		–		–		–		–		–		–		–		–
Coating, bakelite lacquer on unknown substrate		–		–	9	1112		–						–		–		–
Coating, barium fluoride on zinc selenide substrate		–		–		–	9	810		–	9	812		–		–		–
Coating, barium + strontium on nickel substrate		–		–	9	616		–		–				–		–		–
Coating, barium titanate on:																		
Aluminum substrate		–		–	9	815		–	9	818		–		–		–		–
Nb-1Zr substrate		–		–	9	815 817		–		–		–		–		–		–
Coating, bismuth oxide on glass substrate		–		–		–	9	820		–	9	823		–		–		–
Coating, black nickel on copper substrate		–		–	9	700		–		–		–		–		–		–
Coating, boron on Nb-1Zr substrate		–		–	9	826		–		–		–		–		–		–
Coating, boron carbide on:																		
Inconel X substrate		–		–	9	829	9	832		–		–		–		–		–
Molybdenum substrate		–		–	9	827		–		–		–		–		–		–
Coating, butyl acrylate on anodized aluminum substrate		–		–		–	9	1114		–		–		–		–		–
Coating, butylated melamine formaldehyde on anodized aluminum substrate		–		–		–	9	1116		–		–		–		–		–
Coating, butylated urea formaldehyde on anodized aluminum substrate		–		–		–	9	1118		–		–		–		–		–
Coating, butylated urea formaldehyde on aluminum substrate		–		–		–	9	1118		–		–		–		–		–
Coating, cadmium arsenide on glass substrate		–		–		–		–		–	9	836		–		–		–
Coating, cadmium sulfide on aluminum and glass substrate		–		–		–		–		–	9	840		–		–		–
Coating, cadmium oxide on glass substrate		–		–		–		–		–	9	838		–		–		–
Coating, calcium on glass substrate		–		–		–		–		–	9	617		–		–		–
Coating, calcium titanate on:																		
Aluminum substrate		–		–	9	845		–	9	850		–		–		–		–
Beryllium substrate		–		–	9	843		–		–		–		–		–		–
Niobium substrate		–		–	9	843 849		–		–		–		–		–		–
Nb-1Zr substrate		–		–	9	843		–		–		–		–		–		–
Stainless steel substrate		–		–	9	843		–		–		–		–		–		–

Substance Name	Thermal Conductivity		Specif. Heat		Thermal Radiative Properties								Thermal Diffusivity		Viscosity		Thermal Expansion	
					Emissivity		Reflectivity		Absorptivity		Transmissiv.							
	V.	Page	V.	Page	V.	Page	V.	Page	V.	Page	V.	Page	V.	Page	V.	Page	V.	Page
Coating, carbon on:																		
Aluminum oxide substrate	–		–		9	854	–		–		–		–		–		–	
Brass substrate	–		–		–		9	859	–		–		–		–		–	
Copper substrate	–		–		9	852 857	9	859	–		–		–		–		–	
Molybdenum substrate	–		–		9	852	–		9	862	–		–		–		–	
Pyroxylin substrate	–		–		–		–		–		9	861	–		–		–	
Silicon dioxide substrate	–		–		–		9	859	–		–		–		–		–	
Silver substrate	–		–		–		9	859	–		–		–		–		–	
Tantalum substrate	–		–		9	854 855	–		–		–		–		–		–	
Coating, carbon dioxide on stainless steel substrate	–		–		–		9	864	–		–		–		–		–	
Coating, cat-A-Lac clear on polyethylene substrate	–		–		–		–		–		9	1122	–		–		–	
Coating, cerium dioxide on:																		
Nimonic 75 substrate	–		–		9	870	–		–		–		–		–		–	
Tungsten substrate	–		–		9	868 872	–		–		–		–		–		–	
Coating, chromium on:																		
Copper substrate	–		–		9	618	–		–		–		–		–		–	
Glass substrate	–		–		–		9	621	–		–		–		–		–	
Monel substrate	–		–		9	618	–		–		–		–		–		–	
Nickel substrate	–		–		–		9	621	9	626 627	–		–		–		–	
Silver substrate	–		–		–		9	621	–		–		–		–		–	
Stainless steel substrate	–		–		9	619	9	621 624	–		–		–		–		–	
Coating, chromium + aluminum oxide + ΣXi on inconel substrate	–		–		9	629	–		–		–		–		–		–	
Coating, chromium black on:																		
Aluminum substrate	–		–		9	1173	–		–		–		–		–		–	
Copper substrate	–		–		9	1173	–		–		–		–		–		–	
Stainless steel substrate	–		–		9	1173	–		–		–		–		–		–	
Coating, chromium carbide + cobalt on Armco iron substrate	–		–		9	873	–		9	874	–		–		–		–	
Coating, chromium oxide on:																		
Aluminum substrate	–		–		9	876	–		–		–		–		–		–	
Niobium substrate	–		–		9	876	–		–		–		–		–		–	
Coating, chromium oxide + silicon dioxide + ΣXi on:																		
Inconel substrate	–		–		9	885	–		–		–		–		–		–	
Molybdenum substrate	–		–		9	878 883	–		–		–		–		–		–	
Niobium substrate	–		–		9	878 885	–		–		–		–		–		–	
Stainless steel substrate	–		–		9	878	–		–		–		–		–		–	
Steel substrate	–		–		9	881 885	–		–		–		–		–		–	

Substance Name	Thermal Conductivity		Specif. Heat		Thermal Radiative Properties								Thermal Diffusivity		Viscosity		Thermal Expansion	
					Emissivity		Reflectivity		Absorptivity		Transmissiv.							
	V.	Page	V.	Page	V.	Page	V.	Page	V.	Page	V.	Page	V.	Page	V.	Page	V.	Page
Coating, chromium oxide + silicon dioxide + ΣXi on: (continued)																		
Titanium 6Al-4V substrate		–		–	9	881 883		–		–		–		–		–		–
Coating, cobalt on:																		
Glass substrate		–		–		–	9	632		–	9	634		–		–		–
Platinum substrate		–		–	9	631		–	9	633		–		–		–		–
Stainless steel substrate		–		–	9	631	9	632		–		–		–		–		–
Coating, cobalt oxide on:																		
Silver substrate		–		–	9	887		–		–		–		–		–		–
Tantalum substrate		–		–	9	887 888		–		–		–		–		–		–
Coating, cobalt + tungsten on inconel X substrate		–		–	9	636	9	639		–		–		–		–		–
Coating, copper on:																		
Epoxy substrate		–		–		–	9	643		–		–		–		–		–
Glass substrate		–		–		–	9	642		–	9	644		–		–		–
Polyurethane substrate		–		–		–	9	643		–		–		–		–		–
Coating, copper oxide on:																		
Nickel substrate		–		–	9	891		–		–		–		–		–		–
Silver substrate		–		–	9	891		–		–		–		–		–		–
Stainless steel substrate		–		–	9	890		–		–		–		–		–		–
Coating, copper + tin on:																		
Glass substrate		–		–		–	9	645		–		–		–		–		–
Steel substrate		–		–		–	9	645		–		–		–		–		–
Coating, copper phosphorous selenide on fluorite substrate		–		–		–	9	892		–	9	893		–		–		–
Coating, copper sulfide black on copper substrate		–		–	9	894		–		–		–		–		–		–
Coating, Corning 7940 on silver substrate		–		–	9	1049	9	1054	9	1072		–		–		–		–
Coating, cymel 405 on quartz substrate		–		–		–		–	9	1130		–		–		–		–
Coating, diacetyl cellulose on varnish substrate		–		–		–		–		–	9	1119		–		–		–
Coating, Dow 7 on magnesium substrate		–		–	9	1274		–		–		–		–		–		–
Coating, Dow 15 on magnesium substrate		–		–	9	1274		–	9	1275		–		–		–		–
Coating, Dow 17 on magnesium substrate		–		–	9	1274		–	9	1275		–		–		–		–
Coating, Dow Corning 6510 on aluminum substrate		–		–		–	9	1159		–		–		–		–		–
Coating, dry ice on cat-A-Lac black substrate		–		–		–	9	864		–		–		–		–		–
Coating, dry ice on stainless steel substrate		–		–		–	9	864		–		–		–		–		–
Coating, Dutch Boy quick drying enamel on aluminum substrate		–		–		–	9	13		–		–		–		–		–
Coating, elvanol on fiberglass substrate		–		–		–	9	1145	9	1146		–		–		–		–

Substance Name	Thermal Conductivity		Specif. Heat		Thermal Radiative Properties								Thermal Diffusivity		Viscosity		Thermal Expansion	
					Emissivity		Reflectivity		Absorptivity		Transmissiv.							
	V.	Page	V.	Page	V.	Page	V.	Page	V.	Page	V.	Page	V.	Page	V.	Page	V.	Page
Coating, epoxy on:																		
Aluminum alloy substrate	–		–		9	1120	–		–		–		–		–		–	
Polyethylene substrate	–		–		–		–		–		9	1122	–		–		–	
Coating, fasson foil on copper substrate	–		–		–		9	594	–		–		–		–		–	
Coating, FCZ-11 on Nb-1Zr substrate	–		–		9	1024 1028	–		–		–		–		–		–	
Coating, flight data of space vehecles;																		
Lunar orbiter I	–		–		–		–		9	394	–		–		–		–	
Lunar orbiter II	–		–		–		–		9	394	–		–		–		–	
Lunar orbiter IV	–		–		–		–		9	281 393	–		–		–		–	
Lunar orbiter V	–		–		–		–		9	100 393 1167	–		–		–		–	
Mariner IV	–		–		–		–		9	394	–		–		–		–	
Mariner V	–		–		–		–		9	394	–		–		–		–	
OSO II	–		–		–		–		9	60 153 205 207 272 281 376 392 441	–		–		–		–	
OSO III	–		–		–		–		9	394 612 1072 1154 1167 1230	–		–		–		–	
OVI-10	–		–		–		–		9	275	–		–		–		–	
Pegasus I	–		–		–		–		9	394	–		–		–		–	
Pegasus II	–		–		–		–		9	394	–		–		–		–	
Coating, germanium on:																		
Calcium fluoride substrate	–		–		–		9	647	–		–		–		–		–	
Lead chloride substrate	–		–		–		9	647	–		–		–		–		–	
Lithium fluoride substrate	–		–		–		9	647	–		9	650	–		–		–	
Coating, glass on aluminum substrate	–		–		–		9	1035	9	1036	–		–		–		–	
Coating, glass on silver substrate	–		–		–		9	1035	–		–		–		–		–	
Coating, glyptal, clear on aluminum substrate	–		–		–		9	1111	–		–		–		–		–	
Coating, gold on:																		
Aluminum substrate	–		–		9	652	–		–		–		–		–		–	
Aluminum alloy substrate	–		–		–		9	666	–		–		–		–		–	
Cerium oxide substrate	–		–		9	656	–		–		–		–		–		–	
Copper substrate	–		–		9	651	–		9	675	–		–		–		–	
Dow 17 substrate	–		–		9	660	–		–		–		–		–		–	
Epoxy substrate	–		–		–		9	672	–		–		–		–		–	
Fiberglass substrate	–		–		9	651 660	–		–		–		–		–		–	
Glass substrate	–		–		9	656	9	664	–		–		–		–		–	

Substance Name	Thermal Conductivity		Specif. Heat		Thermal Radiative Properties								Thermal Diffusivity		Viscosity		Thermal Expansion	
					Emissivity		Reflectivity		Absorptivity		Transmissiv.							
	V.	Page	V.	Page	V.	Page	V.	Page	V.	Page	V.	Page	V.	Page	V.	Page	V.	Page
Coating, gold on: (continued)																		
Inconel X substrate		–		–	9	656		–		–		–		–		–		–
Magnesium substrate		–		–	9	660		–		–		–		–		–		–
Molybdenum substrate		–		–	9	651		–		–		–		–		–		–
Mylar substrate		–		–	9	651		–	9	675		–		–		–		–
NBS ceramic A418 substrate		–		–	9	656		–		–		–		–		–		–
Nickel substrate		–		–	9	652	9	672		–		–		–		–		–
Nickel oxide substrate		–		–	9	656		–		–		–		–		–		–
Polyester substrate		–		–	9	651		–		–		–		–		–		–
Polyurethane substrate		–		–		–	9	672		–		–		–		–		–
Quartz substrate		–		–		–	9	664	9	678	9	681		–		–		–
Silicone substrate		–		–		–	9	672		–		–		–		–		–
Stainless steel substrate		–		–	9	651	9	673	9	675		–		–		–		–
Titanium substrate		–		–	9	660	9	665		–		–		–		–		–
Coating, gold + palladium + ΣXi on glass substrate		–		–		–	9	683		–		–		–		–		–
Coating, gold + silver on:																		
Copper substrate		–		–	9	651		–	9	675		–		–		–		–
Stainless steel substrate		–		–	9	651		–	9	675		–		–		–		–
Coating, gold black on:																		
Cellulose nitrate substrate		–		–		–	9	1175	9	1178	9	1181		–		–		–
Brass substrate		–		–		–	9	1175		–		–		–		–		–
Glass substrate		–		–		–	9	1175		–	9	1182		–		–		–
Coating, gold black + copper black on sodium chloride substrate		–		–		–		–		–	9	1185		–		–		–
Coating, gold black + nickel black on sodium chloride substrate		–		–		–		–		–	9	1187		–		–		–
Coating, graphite on:																		
Aluminum oxide substrate		–		–	9	854		–		–		–		–		–		–
Brass substrate		–		–		–	9	859		–		–		–		–		–
Copper substrate		–		–	9	857		–		–		–		–		–		–
Silicon dioxide substrate		–		–		–	9	859		–		–		–		–		–
Tantalum substrate		–		–	9	854 855		–		–		–		–		–		–
Coating, hafnium oxide on:																		
Quartz, fused, substrate		–		–		–		–		–	9	897		–		–		–
Tungsten substrate		–		–	9	895 896		–		–		–		–		–		–
Coating, hanovia liquid gold on:																		
Ceramic tile substrate		–		–		–	9	672		–		–		–		–		–
Glass substrate		–		–	9	656	9	664		–		–		–		–		–
Inconel X substrate		–		–	9	656		–		–		–		–		–		–
Molybdenum substrate		–		–	9	651		–		–		–		–		–		–
Titanium substrate		–		–	9	660	9	665		–		–		–		–		–

Substance Name	Thermal Conductivity		Specif. Heat		Thermal Radiative Properties								Thermal Diffusivity		Viscosity		Thermal Expansion	
					Emissivity		Reflectivity		Absorptivity		Transmissiv.							
	V.	Page	V.	Page	V.	Page	V.	Page	V.	Page	V.	Page	V.	Page	V.	Page	V.	Page
Coating, hanovia liquid palladium on glass substrate	–		–			–	9	683	–			–		–		–		–
Coating, hanovia liquid platinum on ceramic tile substrate	–		–			–	9	724	–			–		–		–		–
Coating, Hastelloy on stainless steel substrate	–		–		9	717		–		–		–		–		–		–
Coating, Haynes LT-1, LT-1B and LT-2 cermets on Inconel substrate	–		–		9	629 774		–		–		–		–		–		–
Coating, Hughes H-2, titanium dioxide pigment in potassium silicate binder	–		–		9	214		–	9	275 281		–		–		–		–
Coating, Hughes H-10, china clay pigment in silicone resin binder	–		–		9	95	9	97	9	98 100		–		–		–		–
Coating, indium arsenide on glass substrate	–		–			–		–		–	9	899		–		–		–
Coating, iridium on glass substrate	–		–			–	9	685		–	9	686		–		–		–
Coating, iron on glass substrate	–		–			–	9	687		–	9	688		–		–		–
Coating, iron oxide on:																		
Aluminum substrate	–		–			–	9	903		–		–		–		–		–
Gold substrate	–		–			–	9	903		–		–		–		–		–
Haynes alloy 25 substrate	–		–		9	900 901		–		–		–		–		–		–
Coating, iron titanate on:																		
Beryllium substrate	–		–		9	905		–		–		–		–		–		–
Nb-1 Zr substrate	–		–		9	905		–		–		–		–		–		–
Stainless steel substrate	–		–		9	905		–		–		–		–		–		–
Coating, iron titanium aluminum oxide on Nb-1Zr substrate	–		–		9	909 911		–		–		–		–		–		–
Coating, jet dry black No. 78 on aluminum substrate	–		–			–	9	519		–		–		–		–		–
Coating, kapton on aluminum substrate	–		–			–	9	1037 1039	9	1041		–		–		–		–
Coating, krylon on aluminum substrate	–		–			–	9	526		–		–		–		–		–
Coating, krylon on anodized aluminum substrate	–		–		9	1109	9	526		–		–		–		–		–
Coating, lacquer on:																		
Aluminum alloy substrate	–		–		9	1124		–	9	1127		–		–		–		–
Copper substrate	–		–			–	9	1125		–		–		–		–		–
Dow metal substrate	–		–		9	1124		–	9	1127		–		–		–		–
Quartz substrate	–		–			–		–	9	1126	9	1129		–		–		–
Steel substrate	–		–			–		–	9	1126		–		–		–		–
Coating, laminar X-500 on:																		
Aluminum substrate	–		–			–	9	1138 1140		–		–		–		–		–
Polyethylene substrate	–		–			–		–		–	9	1143		–		–		–
Coating, lanthanum antimonide on glass substrate	–		–			–	9	689		–		–		–		–		–

Substance Name	Thermal Conductivity		Specif. Heat		Thermal Radiative Properties								Thermal Diffusivity		Viscosity		Thermal Expansion	
					Emissivity		Reflectivity		Absorptivity		Transmissiv.							
	V.	Page	V.	Page	V.	Page	V.	Page	V.	Page	V.	Page	V.	Page	V.	Page	V.	Page
Coating, lead chloride on germanium substrate	–		–			–	9	912		–		–		–		–		–
Coating, lead molybdenum tetraoxide on:																		
Glass substrate	–		–			–	9	913		–		–		–		–		–
Potassium bromide substrate	–		–			–	9	913		–		–		–		–		–
Coating, lead + tin on copper substrate	–		–		9	690		–	9	691		–		–		–		–
Coating, liquid platinum on:																		
Ceramic tile substrate	–		–			–	9	724		–		–		–		–		–
Glass substrate	–		–		9	722		–		–		–		–		–		–
Quartz substrate	–		–			–	9	724		–		–		–		–		–
Coating, lithium fluoride on:																		
Aluminum substrate	–		–			–	9	1090		–		–		–		–		–
Glass substrate	–		–			–	9	1090		–		–		–		–		–
KRS-5 substrate	–		–			–		–		–	9	1092		–		–		–
Coating, magnesium + aluminum on glass substrate	–		–			–	9	693		–		–		–		–		–
Coating, magnesium aluminate on:																		
Aluminum substrate	–		–		9	915		–	9	918		–		–		–		–
Nb-1Zr substrate	–		–		9	915 917		–		–		–		–		–		–
Coating, magnesium fluoride on:																		
Glass substrate	–		–			–	9	1096		–	9	1105		–		–		–
Iron oxide substrate	–		–		9	1093	9	1096	9	1103		–		–		–		–
Platinum substrate	–		–			–	9	1098	9	1101	9	1105		–		–		–
Quartz, fused, substrate	–		–			–	9	1096 1098	9	1101	9	1105		–		–		–
Silicon dioxide substrate	–		–			–	9	1096 1098	9	1101	9	1105		–		–		–
Silicon monoxide substrate	–		–		9	1094	9	1096		–		–		–		–		–
Coating, magnesium on glass substrate	–		–			–	9	692		–		–		–		–		–
Coating, magnesium oxide on:																		
Aluminum substrate	–		–			–	9	924		–		–		–		–		–
Black paint substrate	–		–			–	9	924		–		–		–		–		–
Nimonic 75 substrate	–		–		9	919		–		–		–		–		–		–
Coating, manganese on glass substrate	–		–			–	9	694		–	9	695		–		–		–
Coating, melamine formaldehyde on quartz substrate	–		–			–		–	9	1130		–		–		–		–
Coating, Metco XP-1103 on Armco iron substrate	–		–		9	696		–	9	697		–		–		–		–
Coating, Metco XP-1106 on Armco iron substrate	–		–		9	764		–	9	772		–		–		–		–
Coating, Metco XP-1109 on Armco iron substrate	–		–		9	873		–	9	874		–		–		–		–

Substance Name	Thermal Conductivity		Specif. Heat		Thermal Radiative Properties								Thermal Diffusivity		Viscosity		Thermal Expansion	
					Emissivity		Reflectivity		Absorptivity		Transmissiv.							
	V.	Page	V.	Page	V.	Page	V.	Page	V.	Page	V.	Page	V.	Page	V.	Page	V.	Page
Coating, Metco XP-1110 on Armco iron substrate	–		–		9	995	–		9	996	–		–		–		–	
Coating, Metco XP-1114 on:																		
Niobium substrate	–		–		9	981 984	–		–		–		–		–		–	
Stainless steel substrate	–		–		9	981	–		–		–		–		–		–	
Coating, Metco XP-1121 on:																		
Nb-1Zr substrate	–		–		9	994	–		–		–		–		–		–	
Stainless steel substrate	–		–		9	994	–		–		–		–		–		–	
Coating, molybdenum conversion	–		–		9	1331 1333 1335 1337	9	1342	9	1345	–		–		–		–	
Coating, molybdenum disilicide on:																		
Bronze substrate	–		–		9	928	–		–		–		–		–		–	
Molybdenum substrate	–		–		9	927 928 929	–		–		–		–		–		–	
VM-1 molybdenum substrate	–		–		9	927	–		–		–		–		–		–	
Coating, molybdenum on Armco iron substrate	–		–		9	696	–		9	697	–		–		–		–	
Coating, molybdenum sulfide on Inconel X substrate	–		–		–		9	930	–		–		–		–		–	
Coating, mylar on aluminum substrate	–		–		9	1042	–		–		–		–		–		–	
Coating, NBS A-418 on Inconel substrate	–		–		9	455 459	–		–		–		–		–		–	
Coating, NBS A-418 on stainless steel substrate	–		–		9	455 459	–		–		–		–		–		–	
Coating, NBS N-143 on Inconel substrate	–		–		9	445 448	–		–		–		–		–		–	
Coating, NBS N-143 on stainless steel substrate	–		–		9	445 448	–		–		–		–		–		–	
Coating, neodymium on fused quartz substrate	–				–		–		–		9	698	–		–		–	
Coating, nickel on:																		
Copper substrate	–		–		9	700	9	702	–		–		–		–		–	
Epoxy substrate	–		–		9	700	9	704	–		–		–		–		–	
Polyurethane substrate	–		–		–		9	704	–		–		–		–		–	
Steel substrate	–		–		9	700	–		–		–		–		–		–	
Coating, nickel aluminide on Inconel substrate	–		–		9	706	–		–		–		–		–		–	
Coating, nickel chromate on Nb-1Zr substrate	–		–		9	932	–		–		–		–		–		–	
Coating, nickel + chromium on Inconel X substrate	–		–		9	709	9	712	–		–		–		–		–	
Coating, nickel + chromium + ΣXi on Hastelloy X substrate	–		–		9	707	–		–		–		–		–		–	
Coating, nickel + cobalt on stainless steel substrate	–		–		9	715	9	716	–		–		–		–		–	
Coating, nickel + molybdenum + ΣXi on stainless steel substrate	–		–		9	717	–		–		–		–		–		–	
Coating, niobium + titanium conversion	–		–		9	1350	–		–		–		–		–		–	

Substance Name	Thermal Conductivity		Specif. Heat		Thermal Radiative Properties								Thermal Diffusivity		Viscosity		Thermal Expansion	
					Emissivity		Reflectivity		Absorptivity		Transmissiv.							
	V.	Page	V.	Page	V.	Page	V.	Page	V.	Page	V.	Page	V.	Page	V.	Page	V.	Page
Coating, niobium + zirconium conversion	–		–		9	1352	–		–		–		–		–		–	
Coating, nitrocellulose on copper substrate			–		9	1131	–		–		–		–		–		–	
Coating, nylon on stainless steel substrate			–			–	9	1132	–		–		–		–		–	
Coating, optical black, Jersey standard			–		9	577	–		–		–		–		–		–	
Coating, Owens-Illinois 650 on aluminum substrate			–			–		–	9	1167		–	–		–		–	
Coating, pack cementation on:																		
Molybdenum substrate			–		9	1333 1337	9	1342	–		–		–		–		–	
Niobium substrate			–		9	1347 1348	9	1349	–		–		–		–		–	
Tantanium substrate			–		9	1353 1355	9	1358	–		–		–		–		–	
Titanium substrate			–		9	1359 1360	9	1362	–		–		–		–		–	
Tungsten substrate			–		9	1363 1364	9	1365	–		–		–		–		–	
Coating, palladium on																		
Glass substrate			–		9	718	9	720		–	9	721	–		–		–	
Inconel X substrate			–		9	718		–		–		–	–		–		–	
Silicon substrate			–			–	9	720		–		–	–		–		–	
Coating, platinum on:																		
Ceramic tile substrate			–			–	9	724		–		–	–		–		–	
Glass substrate			–		9	722		–		–		–	–		–		–	
Quartz, fused, substrate			–			–	9	724	9	727	9	728	–		–		–	
Coating, platinum black on unknown substrate			–			–	9	1188		–		–	–		–		–	
Coating, polybutadiene on tin oxide substrate			–			–		–		–	9	1134	–		–		–	
Coating, polyester on:																		
Aluminum substrate			–		9	1135		–		–		–	–		–		–	
Gold substrate			–		9	1135		–		–		–	–		–		–	
Coating, polymide fluorinated ethylene propylene on silver substrate			–		9	1133		–		–		–	–		–		–	
Coating, polystyrene on glass substrate			–			–	9	1136 1137		–		–	–		–		–	
Coating, polyurethane on:																		
Aluminum substrate			–			–	9	1138 1140		–		–	–		–		–	
Polyethylene substrate			–			–		–		–	9	1143	–		–		–	
Stainless steel substrate			–			–	9	1140		–		–	–		–		–	
Coating, polyvinyl alcohol on fiberglass substrate			–			–	9	1145	9	1146		–	–		–		–	
Coating, polyvinyl butyral on quartz substrate			–			–		–	9	1147		–	–		–		–	
Coating, polyvinyl chloride on:																		
Aluminum substrate			–			–	9	1150	9	1154		–	–		–		–	
Fiberglass substrate			–			–	9	1150	9	1152		–	–		–		–	

Substance Name	Thermal Conductivity		Specif. Heat		Thermal Radiative Properties								Thermal Diffusivity		Viscosity		Thermal Expansion	
					Emissivity		Reflectivity		Absorptivity		Transmissiv.							
	V.	Page	V.	Page	V.	Page	V.	Page	V.	Page	V.	Page	V.	Page	V.	Page	V.	Page
Coating, polyvinyl chloride on: (continued)																		
Nylon substrate	–		–		–		9	1150	9	1152	–		–		–		–	
Coating, polyvinylidene chloride on silicon substrate	–		–		–		–		–		9	1155	–		–		–	
Coating, potassium bromide on platinum substrate	–		–		9	934	–		–		–		–		–		–	
Coating, potassium chloride on lithium fluoride substrate	–		–		–		–		–		9	935	–		–		–	
Coating, potassium iodide on lithium fluoride substrate	–		–		–		9	936	–		–		–		–		–	
Coating, potassium silicate on:																		
Aluminum substrate	–		–		–		9	937 1044	–		–		–		–		–	
Quartz substrate	–		–		–		–		9	938	–		–		–		–	
Coating, potassium tantalate on																		
Gold substrate	–		–		–		9	1045	–		–		–		–		–	
Platinum substrate	–		–		–		9	1045	–		–		–		–		–	
Coating, PRF-6 $MoSi_2$ on molybdenum substrate	–		–		9	929 1331	–		–		–		–		–		–	
Coating, Rhodium on:																		
Inconel X substrate	–		–		9	730	–		–		–		–		–		–	
Stainless steel substrate	–		–		9	729	–		9	731	–		–		–		–	
Coating, rokide A on:																		
Inconel, oxidized, substrate	–		–		9	792	–		–		–		–		–		–	
Molybdenum substrate	–		–		9	788	9	796	9	803	–		–		–		–	
Stainless steel substrate	–		–		9	788 790	9	796	9	805	–		–		–		–	
Coating, rokide C on:																		
Inconel substrate	–		–		9	885	–		–		–		–		–		–	
Molybdenum substrate	–		–		9	878 883	–		–		–		–		–		–	
Niobium substrate	–		–		9	878 885	–		–		–		–		–		–	
Stainless steel substrate	–		–		9	878	–		–		–		–		–		–	
Steel substrate	–		–		9	881 885	–		–		–		–		–		–	
Titanium 6Al-4V substrate	–		–		9	881 883	–		–		–		–		–		–	
Coating, rokide MA on:																		
Aluminum substrate	–		–		9	915	–		9	918	–		–		–		–	
Nb-1Zr substrate	–		–		9	915 917	–		–		–		–		–		–	
Coating, rokide ZS on:																		
Aluminum substrate	–		–		9	1024	–		9	1030	–		–		–		–	
Nb-1Zr substrate	–		–		9	1024	–		–		–		–		–		–	
Stainless steel substrate	–		–		9	1024 1026	9	1029	–		–		–		–		–	
Coating, RTV-602 on aluminum substrate	–		–		9	1156	9	1159	9	1164 1167	–		–		–		–	

Substance Name	Thermal Conductivity V.	Page	Specif. Heat V.	Page	Emissivity V.	Page	Reflectivity V.	Page	Absorptivity V.	Page	Transmissiv. V.	Page	Thermal Diffusivity V.	Page	Viscosity V.	Page	Thermal Expansion V.	Page
Coating, rokide Z on 321 stainless steel substrate	-		-		9	1011	9	1018	-		-		-		-		-	
Coating, rubidium iodide on lithium fluoride substrate	-		-		-		9	940	-		-		-		-		-	
Coating, SR-111 on silver substrate	-		-		-		9	1163	-		-		-		-		-	
Coating, SY-627-119 on stainless steel substrate	-		-		-		9	1140	-		-		-		-		-	
Coating, sapphire on:																		
Aluminum substrate	-		-		9	785	9	794 796 800	9	803	-		-		-		-	
Gold substrate	-		-		-		-		9	803	-		-		-		-	
Silver substrate	-		-		9	785 788	9	796	9	802 803	-		-		-		-	
Stainless steel substrate	-		-		-		-		9	804	-		-		-		-	
Coating, selenium on:																		
Germanium substrate	-		-		-		-		-		9	943	-		-		-	
Pliofilm substrate	-		-		-		-		-		9	943	-		-		-	
Silicon dioxide substrate	-		-		-		-		-		9	943	-		-		-	
Coating, silicon on:																		
Aluminum substrate	-		-		9	1156	9	1159	9	1164 1167	-		-		-		-	
Dow metal substrate	-		-		9	1156	-		9	1164	-		-		-		-	
Mild steel substrate	-		-		9	1156	-		9	1164	-		-		-		-	
Silver substrate	-		-		-		9	1163	-		-		-		-		-	
Stainless steel substrate	-		-		9	1156	-		9	1164	-		-		-		-	
Coating, silicon carbide on:																		
Aluminum oxide substrate	-		-		9	949	-		-		-		-		-		-	
Graphite substrate	-		-		9	945 947	9	950	-		-		-		-		-	
Tantalum substrate	-		-		9	945 947	-		-		-		-		-		-	
Coating, silicon dioxide on:																		
Aluminum substrate	-		-		9	1049 1052	9	1054 1057 1061	9	1068 1070	-		-		-		-	
Gold substrate	-		-		-		9	1060	9	1068	-		-		-		-	
Magnesium substrate	-		-		9	1049	-		9	1070	-		-		-		-	
Nickel substrate	-		-		-		9	1060	-		-		-		-		-	
Nimonic 75 substrate	-		-		9	1052	-		-		-		-		-		-	
Niobium substrate	-		-		9	1049	-		-		-		-		-		-	
Platinum substrate	-		-		-		9	1061	9	1067	9	1073	-		-		-	
Silicon substrate	-		-		-		-		-		9	1073	-		-		-	
Silver substrate	-		-		9	1049 1052	9	1054 1060 1061	9	1065 1068 1070 1072	-		-		-		-	
Taylor wire substrate	-		-		9	1049	-		-		-		-		-		-	
Coating, silicon Durak-B on molybdenum substrate	-		-		9	1331	-		-		-		-		-		-	

Substance Name	Thermal Conduc- tivity		Specif. Heat		Thermal Radiative Properties								Thermal Diffu- sivity		Visco- sity		Thermal Expan- sion	
					Emis- sivity		Reflec- tivity		Absorp- tivity		Trans- missiv.							
	V.	Page	V.	Page	V.	Page	V.	Page	V.	Page	V.	Page	V.	Page	V.	Page	V.	Page
Coating, silicon Durak—MG on molybdenum substrate	–		–		9	1333 1335	–		–		–		–		–		–	
Coating, silicon monoxide on:																		
Aluminum substrate	–		–		9	1074	9	1077	9	1080	–		–		–		–	
Aluminum oxide substrate	–		–		–		9	1077	–		–		–		–		–	
Inconel substrate	–		–		9	1075	–		–		–		–		–		–	
Platinum substrate	–		–		9	1075	–		–		–		–		–		–	
Coating, silicon nitride on:																		
Gallium arsenide substrate	–		–				–		–		–		9	953	–		–		–	
Silicon substrate	–		–				–		–		–		9	953	–		–		–	
Coating, silicone binder with TiO$_2$ on Dow 15 treated Mg alloy substrate	–		–		9	212	–		9	263	–		–		–		–	
Coating, silver on:																		
Chromium substrate	–		–		–		9	738	–		–		–		–		–	
Copper substrate	–		–		9	734	–		9	747	–		–		–		–	
Epoxy substrate	–		–		9	733	9	742	–		–		–		–		–	
Glass substrate	–		–		–		9	738 743	–		9	752	–		–		–	
Mylar substrate	–		–		–		9	746	–		9	750	–		–		–	
Nickel substrate	–		–		9	734	–		9	747	–		–		–		–	
Polyurethane substrate	–		–		9	733	9	741	–		–		–		–		–	
Quartz substrate	–		–		–		9	738 743	–		9	750 752	–		–		–	
Sapphire substrate	–		–		–		–		–		9	750	–		–		–	
Silicone substrate	–		–		9	733	9	741	–		–		–		–		–	
Stainless steel substrate	–		–		9	734 736	9	738 742	9	747	–		–		–		–	
Coating, silver + aluminum on glass substrate	–		–		–		9	754	–		–		–		–		–	
Coating, silver black on unknown substrate	–		–		–		–		–		9	1189	–		–		–	
Coating, silver sulfide on silver substrate	–		–		9	954	–		9	955	–		–		–		–	
Coating, sodium chloride on aluminum substrate	–		–		–		9	1081	–		–		–		–		–	
Coating, sodium salicylate on MgO pigmented paint substrate	–		–		–		9	956	–		–		–		–		–	
Coating, solder on copper substrate	–		–		9	758	–		9	759	–		–		–		–	
Coating, speculum on:																		
Glass substrate	–		–		–		9	645	–		–		–		–		–	
Steel substrate	–		–		–		9	645 757	–		–		–		–		–	
Coating, strontium titanate on:																		
Aluminum substrate	–		–		9	958	–		9	963	–		–		–		–	
Nb—1Zr substrate	–		–		9	958 962	–		–		–		–		–		–	
Stainless steel substrate	–		–		9	958	–		–		–		–		–		–	

Substance Name	Thermal Conductivity		Specif. Heat		Thermal Radiative Properties										Thermal Diffusivity		Viscosity		Thermal Expansion	
					Emissivity		Reflectivity		Absorptivity		Transmissiv.									
	V.	Page	V.	Page	V.	Page	V.	Page	V.	Page	V.	Page	V.	Page	V.	Page	V.	Page	V.	Page
Coating, synar on niobium substrate	–		–		9	1049	–		–		–		–		–		–		–	
Coating, T-40-C-C-9 on aluminum substrate	–		–		–		9	1110	–		–		–		–		–	
Coating, TAM-CP on stainless steel substrate	–		–		–		9	1017	–		–		–		–		–	
Coating, tantalum carbide on inconel X substrate	–		–		9	965	9	968	–		–		–		–		–	
Coating, teflon on:																				
Aluminum substrate	–		–		–		9	1085	9	1087	–		–		–		–	
S-13 substrate	–		–		–		9	1085	9	1087	–		–		–		–	
Coating, tellurium on:																				
Glass substrate	–		–		–		9	971	–		–		–		–		–	
Stilbene substrate	–		–		–		–		–		9	972	–		–		–	
Coating, tesslar on aluminum substrate	–		–		–		–		9	1082	–		–		–		–	
Coating, thorium dioxide on:																				
Nimonic 75 substrate	–		–		9	973	–		–		–		–		–		–	
Quartz substrate	–		–		–		–		–		9	974	–		–		–	
Coating, titanium dioxide on:																				
Aluminum substrate	–		–		9	981	9	986 989 990	9	991	9	992	–		–		–	
Black paint substrate	–		–		–		9	986	–		–		–		–		–	
Iron substrate	–		–		9	981	–		–		–		–		–		–	
Nickel substrate	–		–		9	981	–		–		–		–		–		–	
Niobium substrate	–		–		9	981 984	–		–		–		–		–		–	
Stainless steel substrate	–		–		9	981	–		–		–		–		–		–	
Coating, tin on copper substrate	–		–		9	755	–		9	756	–		–		–		–	
Coating, tin + copper on steel substrate	–		–		–		9	757	–		–		–		–		–	
Coating, tin + lead on copper substrate	–		–		9	758	–		9	759	–		–		–		–	
Coating, tin oxide on:																				
Glass substrate	–		–		–		–		–		9	979	–		–		–	
Tin oxide substrate	–		–		–		9	977	–		–		–		–		–	
Coating, titanium on aluminum substrate	–		–		–		9	760	9	763	–		–		–		–	
Coating, titanium on brass substrate	–		–		–		–		9	762	–		–		–		–	
Coating, titanium on glass substrate	–		–		–		9	761	–		–		–		–		–	
Coating, titanium dioxide + aluminum on:																				
Nb-1Z substrate	–		–		9	994	–		–		–		–		–		–	
Stainless steel substrate	–		–		9	994	–		–		–		–		–		–	
Coating, thorium dioxide on nimonic 75 substrate	–		–		9	973	–		–		–		–		–		–	
Coating, tungsten on:																				
Armco iron substrate	–		–		9	764	–		9	772	–		–		–		–	
Inconel X substrate	–		–		9	766	9	769	–		–		–		–		–	

Substance Name	Thermal Conductivity		Specif. Heat		Thermal Radiative Properties								Thermal Diffusivity		Viscosity		Thermal Expansion	
					Emissivity		Reflectivity		Absorptivity		Transmissiv.							
	V.	Page	V.	Page	V.	Page	V.	Page	V.	Page	V.	Page	V.	Page	V.	Page	V.	Page
Coating, tungsten carbide + cobalt on iron substrate	–		–		9	995		–	9	996		–		–		–		–
Coating, tungsten + chromium + aluminum oxide on inconel substrate	–		–		9	774		–		–		–		–		–		–
Coating, tungsten + cobalt on inconel X substrate	–		–			–	9	776		–		–		–		–		–
Coating, uranium on glass substrate	–		–			–	9	779		–	9	780		–		–		–
Coating, uranium dioxide on tungsten substrate	–		–		9	997		–		–		–		–		–		–
Coating, vanadium oxide on:																		
Sapphire substrate	–		–			–	9	999		–	9	1000		–		–		–
Tungsten substrate	–		–		9	998		–		–		–		–		–		–
Coating, vanadium oxide on tungsten substrate	–		–		9	998		–		–		–		–		–		–
Coating, zinc black on:																		
Brass substrate	–		–			–	9	1190		–		–		–		–		–
Metastyrene substrate	–		–			–	9	1190		–		–		–		–		–
Pliofilm substrate	–		–			–		–		–	9	1191		–		–		–
Ptroxylin substrate	–		–			–		–		–	9	1191		–		–		–
Coating, zinc on iron substrate	–		–			–	9	781		–		–		–		–		–
Coating, zinc oxide on titanium substrate	–		–			–	9	1003		–		–		–		–		–
Coating, zinc selenide on quartz substrate	–		–			–	9	1004		–	9	1005		–		–		–
Coating, zinc sulfide on glass substrate	–		–			–	9	1007 1008		–		–		–		–		–
Coating, zinc telluride on quartz substrate	–		–			–		–		–	9	1009		–		–		–
Coating, zirconium on molybdenum substrate	–		–		9	782		–		–		–		–		–		–
Coating, zirconium oxide on:																		
Aluminum substrate	–		–			–	9	1017		–		–		–		–		–
Inconel substrate	–		–		9	1011	9	1017	9	1021		–		–		–		–
Inconel X substrate	–		–		9	1014	9	1017		–		–		–		–		–
Mild steel substrate	–		–		9	1011 1014		–		–		–		–		–		–
Nimonic 75 substrate	–		–		9	1011		–		–		–		–		–		–
Stainless steel substrate	–		–		9	1011	9	1017		–		–		–		–		–
Coating, zirconium silicate on:																		
Aluminum substrate	–		–		9	1024		–	9	1030		–		–		–		–
Nb-1 Zr substrate	–		–		9	1024 1028		–		–		–		–		–		–
Stainless steel substrate	–		–		9	1024 1026	9	1029		–		–		–		–		–
Coating, zirconium titanate on beryllium substrate	–		–		9	1032		–		–		–		–		–		–
Coating, wulfenite on potassium bromide substrate	–		–			–	9	913		–		–		–		–		–
Coating, yttrium oxide on tungsten substrate	–		–		9	1001		–		–		–		–		–		–

Substance Name	Thermal Conductivity V.	Page	Specif. Heat V.	Page	Emissivity V.	Page	Reflectivity V.	Page	Absorptivity V.	Page	Transmissiv. V.	Page	Thermal Diffusivity V.	Page	Viscosity V.	Page	Thermal Expansion V.	Page
Cobalt, Co	1	64	4	48	7	123 126	7	132	–		–		10	48	–		12	70
Cobalt, electrolytic	–		–		–		–		–		–		–		–		12	73
Cobalt alloys:																		
Co + Al	–		–				7	901	–		–		–		–		–	
Co + Au	–		–		–		–		–		–		–		–		12	735
Co + C	1	526	–		–		–		–		–		–		–		–	
Co + Cr	1	527	–		–		–		–		–		–		–		12	705
Co + Dy	–		4	314	–		–		–		–		–		–		–	
Co + Fe	–		4	317	7	904	–		–		–		–		–		12	737
Co + Mn	–		–		–		–		–		–		–		–		12	744
Co + Mo	–		–		–		–		–		–		–		–		12	743
Co + Nb	–		–		–		–		–		–		–		–		12	751
Co + Ni	1	528	4	320	7	906	–		–		–		–		–		12	747
Co + Pd	–		–		–		–		–		–		–		–		12	747
Co + Pt	–		–		–		–		–		–		–		–		12	754
Co + Si	–		–		–		–		–		–		10	229	–		–	
Co + Ti	–		–		–		–		–		–		–		–		12	756
Co + V	–		–		–		–		–		–		–		–		12	761
Co + W	–		–		–		–		–		–		–		–		12	757
Co + Ta	–		–		–		–		–		–		–		–		12	755
Co + Zr	–		–		–		–		–		–		–		–		12	762
Co + Cr + ΣXi	1	947	4	523	7	1138 1142 1145 1148	7	1152	–		–		10	287	–		12	1067
Co + Cr + ΣXi, elgiloy	–		–		–		–		–		–		–		–		12	1067 1069
Co + Cr + ΣXi, British, C-32	1	948	–		–		–		–		–		–		–		–	
Co + Cr + ΣXi, Haynes alloy No. 25	–		–		7	1140 1143 1146 1149	7	1154	–		–		–		–		–	
Co + Cr + ΣXi, Haynes stellite 21	1	948	–		–		–		–		–		–		–		–	
Co + Cr + ΣXi, Haynes stellite 23	1	948	–		–		–		–		–		–		–		–	
Co + Cr + ΣXi, Haynes stellite 25	–		–		–		–		–		–		10	287	–		–	
Co + Cr + ΣXi, Haynes stellite 31	1	948	–		–		–		–		–		–		–		–	
Co + Cr + ΣXi, Haynes stellite HE 1049	–		4	526	–		–		–		–		–		–		12	1070
Co + Cr + ΣXi, S 816	1	948	–		–		–		–		–		–		–		–	
Co + Cr + ΣXi, stellite	–		–		–		7	1154	–		–		–		–		–	
Co + Cr + ΣXi, stellite 3	–		–		–		–		–		–		–		–		12	1067 1069
Co + Cr + ΣXi, stellite 6	–		–		–		–		–		–		–		–		12	1067 1069
Co + Cr + ΣXi, stellite 21	–		4	524	–		–		–		–		–		–		12	1067 1069
Co + Cr + ΣXi, stellite 23	–		–		–		–		–		–		–		–		12	1067 1069

Substance Name	Thermal Conductivity		Specif. Heat		Thermal Radiative Properties								Thermal Diffusivity		Viscosity		Thermal Expansion	
					Emissivity		Reflectivity		Absorptivity		Transmissiv.							
	V.	Page	V.	Page	V.	Page	V.	Page	V.	Page	V.	Page	V.	Page	V.	Page	V.	Page
Cobalt alloys: (continued)																		
Co + Cr + ΣXi, stellite 25		–		–		–		–		–		–		–		–	12	1067 1069
Co + Cr + ΣXi, stellite 27		–		–		–		–		–		–		–		–	12	1067 1069
Co + Cr + ΣXi, stellite 30		–		–		–		–		–		–		–		–	12	1067 1070
Co + Cr + ΣXi, stellite 31		–		–		–		–		–		–		–		–	12	1067 1070
Co + Cr + ΣXi, stellite HE 1049		–	4	526		–		–		–		–		–		–		–
Co + Cr + ΣXi, X-40	1	948		–		–		–		–		–		–		–		–
Co + Cr + ΣXi, WI 52	1	948		–		–		–		–		–		–		–		–
Co + Fe + ΣXi	1	950		–		–		–		–		–		–		–	12	1073
Co + Ni + ΣXi	1	951		–		–		–		–		–		–		–	12	1085
Co + Ni + ΣXi, Nivco alloy		–		–		–		–		–		–		–		–	12	1085 1087
Cobalt-aluminum intermetallic compound, CoAl		–		–		–	8	1352		–		–		–		–		–
Cobalt boride, Co_2B		–		–	8	731		–		–		–		–		–	13	796
Cobalt carbonate, $CoCO_3$		–		–		–		–		–		–		–		–	13	641
Cobalt chlorides:																		
$CoCl_2$		–	5	806		–		–		–		–		–		–		–
$CoCl_2 \cdot 6H_2O$		–	5	809		–		–		–		–		–		–	13	1013
Cobalt-dysprosium intermetallic compound, Co_2Dy		–		–		–		–		–		–		–		–	12	508 510 511
Cobalt fluoride, CoF_2		–	5	934		–		–		–		–		–		–	13	1076
Cobalt-gadolinium intermetallic compound, Co_2Gd		–		–		–		–		–		–		–		–	12	509 510 512
Cobalt iron oxide, $CoO \cdot Fe_2O_3$		–	5	1425		–		–		–		–		–		–		–
Cobalt iron oxide, nonstoichiometric		–	5	1428		–		–		–		–		–		–		–
Cobalt oxides:																		
CoO		–	5	70	8	238 240	8	242		–		–		–		–	13	221
Co_3O_4		–	5	73		–		–		–		–		–		–		–
Cobalt phosphate, $Co_3(PO_4)_2$		–		–		–		–		–		–		–		–	13	689
Cobalt silicate, Co_2SiO_4		–		–		–		–		–		–		–		–	13	727
Cobalt silicides:																		
CoSi		–	5	571	8	1172		–		–		–		–		–	13	1211
Co_3Si		–		–		–		–		–		–		–		–	13	1211
Cobalt strontium titanate, Co_3SrTiO_3	2	271		–		–		–		–		–		–		–		–
Cobalt sulfate heptahydrate, $CoSO_4 \cdot 7H_2O$		–	5	1194		–		–		–		–		–		–		–
Cobalt titanium oxide, $CoO \cdot TiO_2$		–		–		–		–		–		–		–		–	13	563
Cobalt tungsten oxide, $CoO \cdot WO_3$		–	5	1431		–		–		–		–		–		–		–

Substance Name	Thermal Conductivity		Specif. Heat		Thermal Radiative Properties								Thermal Diffusivity		Viscosity		Thermal Expansion	
					Emissivity		Reflectivity		Absorptivity		Transmissiv.							
	V.	Page	V.	Page	V.	Page	V.	Page	V.	Page	V.	Page	V.	Page	V.	Page	V.	Page
Cobalt-yttrium intermetallic compounds:																		
Co_5Y	–		–		–		–		–		–		–		–		12	514 517
$Co_{17}Y_2$	–		–		–		–		–		–		–		–		12	513 515 516
Cobalt zinc ferrate, $Co(Zn)Fe_2O_4$	2	272	–		–		–		–		–		–		–			–
Codfish, pulp	–		–		–		–		–		–		10	635	–			–
Coke, petroleum	2	765	–		–		–		–		–		10	22 35	–			–
Columbium	1	245	4	153	–		–		–		–		10	125	–			–
Cominco 69	–		–		–		–		–		–		–		–		12	395
Composite, asbestos-phenolic 9526 D laminate	–		–		–		–		–		–		–		–		13	1525
Composite, asbestos-phenolic resin	–		–		–		–		–		–		–		–		13	1524
Composite, asbestos reinforced phenolic	–		–		–		–		–		–		–		–		13	1525
Composite, asbestos-teflon	–		–		–		–		–		–		–		–		13	1530
Composite, asphalt-glass wool pad	2	1108	–		–		–		–		–		–		–			–
Composite, boron fiber/epoxy resin	–		–		–		–		–		–		–		–		13	1531
Composite, carbitex 100	–		–		–		–		–		–		–		–		13	1578
Composite, carbitex 700	–		–		–		–		–		–		–		–		13	1578
Composite, graphite fiber/epoxy resin, courtaulds HMS, hercu	–		–		–		–		–		–		–		–		13	1584
Composite, graphite fiber/epoxy resin, courtaulds HMS, pseud	–		–		–		–		–		–		–		–		13	1584
Composite, glass fabric/epoxy resin	–		–		–		–		–		–		–		–		13	1537
Composite, glass fabric/polyester resin	–		–		–		–		–		–		–		–		13	1542
Composite, glass fiber board	2	1124	–		–		–		–		–		–		–			–
Composite, glass fiber/epoxy	–		–		–		–		–		–		–		–		13	1547
Composite, glass fiber/phenolic resin	–		–		–		–		–		–		–		–		13	1559
Composite, glass fiber/phenyl silane resin	–		–		–		–		–		–		–		–		13	1568
Composite, glass fiber reinforced phenolic	–		–		–		–		–		–		–		–		13	1560
Composite, glass fiber/silicone resin	–		–		–		–		–		–		–		–		13	1567
Composite, glass cloth-reinforced/ phenolic resin laminates	–		–		–		–		–		–		–		10	558 559	–			–
Composite, graphite-cloth laminate	–		–		–		–		–		–		–		–		13	1569
Composite, graphite fabric/carbon	–		–		–		–		–		–		–		–		13	1576
Composite, graphite fabric/phenolic resin	–		–		–		–		–		–		–		–		13	1568
Composite, graphite fiber/epoxy resin	–		–		–		–		–		–		–		–		13	1582
Composite, graphite fiber/epoxy resin, modmor II-Narmco 52	–		–		–		–		–		–		–		–		13	1585
Composite, insurok, nonmetallic laminate	2	1023 1024	–		–		–		–		–		–		–			–
Composite, lamicoid, laminate	2	1023 1024	–		–		–		–		–		10	555	–			–
Composite, laminate, epoxy resin	2	1029	–		–		–		–		–		–		–			–
Composite, laminates, metallic-nonmetallic	2	1036	–		–		–		–		–		10	553	–			

Substance Name	Thermal Conduc- tivity		Specif. Heat		Thermal Radiative Properties								Thermal Diffu- sivity		Visco- sity		Thermal Expan- sion	
					Emis- sivity		Reflec- tivity		Absorp- tivity		Trans- missiv.							
	V.	Page	V.	Page	V.	Page	V.	Page	V.	Page	V.	Page	V.	Page	V.	Page	V.	Page
Composite, laminates, nonmetallic		–		–		–		–		–		–	10	554		–		–
Composite, nylon fabric/phenolic resin		–		–		–		–		–		–		–		–	13	1600
Composite, OTWR		–		–		–		–		–		–		–		–	13	1612
Composite, phenolic-asbestos laminate		–		–		–		–		–		–	10	555		–		–
Composite, phenolic-asbestos cloth laminate		–		–		–		–		–		–	10	556 557		–		–
Composite, phenolic-graphite cloth laminate		–		–		–		–		–		–	10	557 558		–		–
Composite, phenolic-graphite mat laminate		–		–		–		–		–		–	10	558		–		–
Composite, phenolic refrasil		–		–		–		–		–		–		–		–	13	1610
Composite, phenolic-refrasil cloth laminate		–		–		–		–		–		–	10	557		–		–
Composite, potassium titanium oxide fiber/epoxy resin		–		–		–		–		–		–		–		–	13	1605
Composite, sandwiches, nonmetallic	2	1044		–		–		–		–		–		–		–		–
Composite, sandwiches, metallic-nonmetallic	2	1047		–		–		–		–		–		–		–		–
Composite, scotchply laminate, nonmetallic	2	1029		–		–		–		–		–		–		–		–
Composite, silicon dioxide fiber/ phenolic resin, RAD-60		–		–		–		–		–		–		–		–	13	1612
Composite, silicon oxide fiber/ phenolic resin		–		–		–		–		–		–		–		–	13	1608
Concrete		–		–		–		–		–		–	10	572		–		–
Concrete, asphaltic bituminous	2	863		–		–		–		–		–		–		–		–
Concrete, baryte		–		–		–		–	8	704		–		–		–		–
Concrete, bitumin	2	863		–		–		–		–		–		–		–		–
Concrete, bituminous aggregate	2	863		–		–		–		–		–		–		–		–
Concrete, cinder aggregate	2	869 870		–		–		–		–		–		–		–		–
Concrete, commercial castable	2	871 875 876 877 878		–		–		–		–		–		–		–		–
Concrete, diatomaceous aggregate	2	874		–		–		–		–		–		–		–		–
Concrete, expanded burned clay aggregate	2	870		–		–		–		–		–		–		–		–
Concrete, foamed light weight	2	881		–		–		–		–		–		–		–		–
Concrete, Haydite aggregate	2	870		–		–		–		–		–		–		–		–
Concrete, light weight	2	874		–		–		–		–		–		–		–		–
Concrete, limestone aggregate	2	869		–		–		–		–		–		–		–		–
Concrete, limestone gravel	2	864 865		–		–		–		–		–		–		–		–
Concrete, lummite cement	2	871		–		–		–		–		–		–		–		–
Concrete, metallurgical pumice	2	863 864		–		–		–		–		–		–		–		–
Concrete, monolithic wall	2	1126		–		–		–		–		–		–		–		–
Concrete, paraffin	2	863		–		–		–		–		–		–		–		–
Concrete, Portland cement	2	871		–		–		–		–		–		–		–		–

Substance Name	Thermal Conductivity		Specif. Heat		Thermal Radiative Properties								Thermal Diffusivity		Viscosity		Thermal Expansion	
					Emissivity		Reflectivity		Absorptivity		Transmissiv.							
	V.	Page	V.	Page	V.	Page	V.	Page	V.	Page	V.	Page	V.	Page	V.	Page	V.	Page
Concrete, sand cement	2	874	–		–		–		–		–		–		–		–	
Concrete, sand and gravel aggregate	2	868 869	–		–		–		–		–		–		–		–	
Concrete, slag	2	864 880 881	–		–		–		–		–		–		–		–	
Concrete, slag aggregate limestone treated	2	870	–		–		–		–		–		–		–		–	
Concrete, slag direct process	2	864	–		–		–		–		–		–		–		–	
Concrete, slag expanded	2	878 879	–		–		–		–		–		–		–		–	
Concrete, slag leuna	2	864	–		–		–		–		–		–		–		–	
Constantan, copper alloy	2	864 880 881	4	341	–		–		–		–		10	234	–		12	781
Copoly(chloroethylene-vinyl-acetate)	2	943	–		–		–		–		–		–		–		–	
Copoly(1,1-difluoroethylene-hexafluoropropene)	–		–		–		–		–		–		–		–		13	1460
Copoly(1,1-difluoroethylene-hexafluoropropene), viton A rubber	2	983	–		–		–		–		–		–		–		–	
Copoly(ethylene-propylene)	–		–		–		–		–		–		–		–		13	1440
Copoly(formaldehyde-urea)	2	944	–		–		–		–		–		–		–		–	
Copoly(formaldehyde-urea), mipora	2	944	–		–		–		–		–		–		–		–	
Copoly(vinyl chloride-vinyl acetate)	–		–		–		–		–		–		–		–		13	1495
Copper, Cu	1	68	4	51	7	136 142 149 152	7	158 165 169 172	7	177 179 181 184 184	–		10	51	–		12	77
Copper, B.S. 1433	–		–		–		7	173 174	7	194	–		–		–		–	
Copper, beryllium	1	539	–		–		–		–		–		–		–		–	
Copper, coalesced	1	69 72	–		–		–		–		–		–		–		–	
Copper, deoxidized	–		–		–		–		–		–		10	58	–		–	
Copper, electrolytic	1	72 73	4	51	–		–		–		–		10	59	–		12	80
Copper, electrolytic tough pitch	1	70 72	4	52	–		–		–		–		10	54	–		–	
Copper, electrolytic tough pitch, QQC 502	–		4	53	–		–		–		–		–		–		–	
Copper, electrolytic tough pitch, QQC 576	–		4	53	–		–		–		–		–		–		–	
Copper, freecutting	1	582	–		–		–		–		–		–		–		–	
Copper, OFHC	1	69 74	4	52	7	138	–		7	189	–		10	53 54 55 57	–		12	80
Copper, OFHC, polycrystalline	–		–		–		–		–		–		–		–		12	80
Copper, phosphorized	–		–		–		–		–		–		10	54	–		–	
Copper, phosphorus deoxidized	1	72	–		–		–		–		–		–		–		–	
Copper, single crystal	–		–		–		–		–		–		–		–		12	80
Copper, spectrographically standardized	–		–		–		–		–		–		10	57	–		–	

Substance Name	Thermal Conductivity		Specif. Heat		Thermal Radiative Properties								Thermal Diffusivity		Viscosity		Thermal Expansion	
					Emissivity		Reflectivity		Absorptivity		Transmissiv.							
	V.	Page	V.	Page	V.	Page	V.	Page	V.	Page	V.	Page	V.	Page	V.	Page	V.	Page
Copper, standard reference material 736		–		–		–		–		–		–		–		–	12	82
Copper alloys:																		
Cu + Ag	1	578		–		–		–		–		–	10	236		–		–
Cu + Al	1	530	4	323		–		–		–		–		–		–		–
Cu + As	1	535		–		–		–		–		–	10	232		–		–
Cu + Au	1	548		–		–		–		–		–		–		–	12	763
Cu + Be	1	538		–		–		–		–		–		–		–	12	678
Cu + Be, beryllium bronze	1	539		–		–		–		–		–		–		–	12	679
Cu + Cd	1	541		–		–		–		–		–		–		–		–
Cu + Co	1	545		–		–		–		–		–		–		–		–
Cu + Cr	1	542		–		–		–		–		–		–		–		–
Cu + Cr, Russian cupralloy, type 5	1	543		–		–		–		–		–		–		–		–
Cu + Fe	1	551	4	331		–		–		–		–		–		–	12	771
Cu + Ga		–	4	327		–		–		–		–		–		–		–
Cu + In		–		–		–		–		–		–		–		–	12	768
Cu + Mg		–	4	335		–		–		–		–		–		–		–
Cu + Mn	1	557	4	338		–		–		–		–		–		–	12	773
Cu + Ni	1	561	4	341	7	908		–		–		–	10	233		–	12	778
Cu + Ni, advance	1	564 970		–		–		–		–		–		–		–		–
Cu + Ni, constantan	1	564	4	341		–		–		–		–	10	234		–	12	781
Cu + Ni, Lohm	1	564		–		–		–		–		–		–		–		–
Cu + Ni, Russian, NM–81 cuprnickel	1	562		–		–		–		–		–		–		–		–
Cu + Ni, Russian cupro nickel, NM–81	1	562		–		–		–		–		–		–		–		–
Cu + P	1	571		–		–		–		–		–		–		–		–
Cu + Pb	1	554		–		–		–		–		–		–		–		–
Cu + Pd	1	568		–		–		–		–		–		–		–		–
Cu + Pt	1	574		–		–		–		–		–		–		–		–
Cu + Sb	1	534		–		–		–		–		–	10	231		–		–
Cu + Si	1	575		–		–		–		–		–		–		–		–
Cu + Sn	1	584		–		–	7	910		–		–	10	237		–	12	788
Cu + Te	1	581		–		–		–		–		–		–		–		–
Cu + Te, ASTM B301–58T	1	582		–		–		–		–		–		–		–		–
Cu + Zn	1	588	4	346	7	912 914	7	917 920	7	923 925 928		–	10	238		–	12	796
Cu + Zn, brass, 70/30	1	590		–		–		–		–		–		–		–		–
Cu + Zn, brass, alpha		–	4	346		–		–		–		–		–		–		–
Cu + Al + ΣXi	1	952		–	7	1159 1162	7	1166	7	1169		–	10	288		–	12	1089
Cu + Al + ΣXi, bronze	1	531 532 953		–	7	1160		–		–		–		–		–	12	1091
Cu + Be + ΣXi	1	955		–		–		–		–		–		–		–	12	1095
Cu + Be + ΣXi, Berylco 25		–		–		–		–		–		–		–		–	12	1097

Substance Name	Thermal Conductivity		Specif. Heat		Thermal Radiative Properties										Thermal Diffusivity		Viscosity		Thermal Expansion	
					Emissivity		Reflectivity		Absorptivity		Transmissiv.									
	V.	Page	V.	Page	V.	Page	V.	Page	V.	Page	V.	Page	V.	Page	V.	Page	V.	Page	V.	Page
Copper alloys: (continued)																				
Cu + Cd + ΣXi	1	956		–		–		–		–		–		–		–				–
Cu + Co + ΣXi	1	957		–		–		–		–		–		–		–				–
Cu + Cr + ΣXi		–	4	526		–		–		–		–		–		–				–
Cu + Fe + ΣXi	1	960		–		–		–		–		–		–		–				–
Cu + Mg + ΣXi		–	4	529		–		–		–		–		–		–				–
Cu + Mn + ΣXi	1	964		–		–		–		–		–		–				–	12	1102
Cu + Mn + ΣXi, manganin	1	965	4	338		–		–		–		–		–		–				–
Cu + Mn + ΣXi, Russian, manganin NM Mts	1	965		–		–		–		–		–		–		–				–
Cu + Ni + ΣXi	1	969		–		–	7	1172		–		–		–				–	12	1103
Cu + Ni + ΣXi, Aterite		–		–		–		–		–		–		–				–	12	1105
Cu + Ni + ΣXi, cupronickel	1	970		–		–		–		–		–		–		–				–
Cu + Ni + ΣXi, Tempaloy 836		–		–		–		–		–		–		–				–	12	1103 1105
Cu + Ni + ΣXi, Tempaloy 841		–		–		–		–		–		–		–				–	12	1089 1091
Cu + Ni + ΣXi, Tempaloy, soft		–		–		–		–		–		–		–				–	12	1105
Cu + Pb + ΣXi	1	961		–		–		–		–		–		–				–	12	1098
Cu + Pb + ΣXi, SAE bearing alloy 40	1	976		–		–		–		–		–		–		–				–
Cu + Pb + ΣXi, SAE bearing alloy 64	1	976		–		–		–		–		–		–		–				–
Cu + Pb + ΣXi, SAE bearing alloy 66	1	962		–		–		–		–		–		–		–				–
Cu + Si + ΣXi	1	972		–		–		–		–		–		–				–	12	1107
Cu + Sn + ΣXi	1	975		–		–	7	1174		–		–		–				–	12	1110
Cu + Sn + ΣXi, gun metal, admiralty	1	976		–		–		–		–		–		–		–				–
Cu + Sn + ΣXi, gun metal, ordinary	1	976		–		–		–		–		–		–		–				–
Cu + Sn + ΣXi, Navy M	1	977		–		–		–		–		–		–		–				–
Cu + Sn + ΣXi, SAE bearing alloy 62	1	976		–		–		–		–		–		–		–				–
Cu + Zn + ΣXi	1	979		–		–		–		–		–		–				–	12	1114
Cu + Zn + ΣXi, German silver	1	980 981		–		–		–		–		–		–				–		–
Cu + Zn + ΣXi, manganese bronze		–		–		–		–		–		–		–				–	12	1116
Cu + Zn + ΣXi, nickel silver	1	981		–		–		–		–		–		–				–	12	1105 1114 1116
Cu + Zr + ΣXi	1	985		–		–		–		–		–		–				–		–
Copper antimony selenide + copper selenide, mixtures	1	1400		–		–		–		–		–		–		–				–
Copper-antimony-selenium intermetallic compound, CuSbSe$_2$	1	1275		–		–		–		–		–		–		–				–
Copper + beryllium cobalt compoumd, mixture	1	1420		–		–		–		–		–		–		–				–
Copper bromide, CuBr		–	5	762		–	8	741		–	8	743		–				–	13	838
Copper chlorides:																				
CuCl		–		–		–	8	856		–	8	858		–				–	13	1013
CuCl$_2$		–	5	812		–		–		–		–		–				–		–
CuCl$_2 \cdot 2H_2O$		–	5	815		–		–		–		–		–				–		–

Substance Name	Thermal Conductivity		Specif. Heat		Thermal Radiative Properties								Thermal Diffusivity		Viscosity		Thermal Expansion	
					Emissivity		Reflectivity		Absorptivity		Transmissiv.							
	V.	Page	V.	Page	V.	Page	V.	Page	V.	Page	V.	Page	V.	Page	V.	Page	V.	Page
Copper glance, copper sulfide	2	699	–		–		–		–		–		–		–		–	
Copper-gold intermetallic compound, CuAu	–		–		–		–		–		–		–		–		12	519 520 522
Copper indium telluride, CuInTe$_2$	–		–		–		–		–		–		–		–		13	1270
Copper iodide, CuI	2	562	–		–		8	999	–		8	1001	–		–		13	1122
Copper iron oxide, CuO·Fe$_2$O$_3$	–		5	1437	–		–		–		–		–		–		–	
Copper iron oxide, nonstoichiometric	–		5	1434	–		–		–		–		–		–		–	
Copper-magnesium intermetallic compound, Cu$_2$Mg	–		–		–		–		–		–		–		–		12	523
Copper oxides:																		
CuO	–		5	80	–		8	247	–		8	249	–		–		–	
Cu$_2$O	2	147	5	76	8	243 245	–		–		8	249	–		–		–	
Copper selenide, Cu$_3$Se$_2$	1	1276	–		–		–		–		–		–		–		–	
Copper selenide-copper antimony selenide, mixture	1	1401	–		–		–		–		–		–		–		–	
Copper silicides:																		
Cu$_5$Si	–		–		–		–		–		–		–		–		13	1211
Cu$_{15}$Si$_4$	–		–		–		–		–		–		–		–		13	1212
Copper sulfides:																		
CuS	–		5	662	–		–		–		–		–		–		–	
Cu$_2$S	2	699	5	665	–		–		–		–		–		–		–	
Copper-tin intermetallic compound, Cu$_4$Sn	–		–		–		8	1352	–		–		–		–		–	
Copper telluride + indium telluride + silver telluride, mixture	1	1406	–		–		–		–		–		–		–		–	
Copper + titanium nickel compound, mixture	1	1433	–		–		–		–		–		–		–		–	
Copper-zinc intermetallic compounds:																		
CuZn	–		–		–		8	1285	–		–		–		–		12	525
Cu$_5$Zn$_6$	–		–		–		–		–		–		–		–		12	524
Copperas	–		5	1200	–		–		–		–		–		–		–	
Coralto-cobaltic oxide	–		5	73	–		–		–		–		–		–		–	
Cordierite	2	918	5	1503	–		–		–		8	1650	–		–		13	727
Cordierite, 202	2	919	–		–		–		–		–		–		–		–	
Cordierite, Rutgers	2	919	–		–		–		–		–		–		–		–	
Cordierite, steatite	2	919	–		–		–		–		–		–		–		–	
Cork	2	1063	–		–		–		–		–		–		–		–	
Corundum	2	94 99	5	26	–		–		–		–		10	383	–		13	179 182 183
Cotton	2	1068	–		–		–		–		–		–		–		–	
Cotton, fabric	2	1093	–		–		–		–		–		–		–		–	
Cotton, medical	2	1069 1070	–		–		–		–		–		–		–		–	

Substance Name	Thermal Conductivity		Specif. Heat		Thermal Radiative Properties								Thermal Diffusivity		Viscosity		Thermal Expansion	
					Emissivity		Reflectivity		Absorptivity		Transmissiv.							
	V.	Page	V.	Page	V.	Page	V.	Page	V.	Page	V.	Page	V.	Page	V.	Page	V.	Page
Cotton, mineral	2	1147	–		–		–		–		–		–		–		–	
Cotton, silicate felt fabric	2	1094	–		–		–		–		–		–		–		–	
Cotton, waste	2	1070	–		–		–		–		–		–		–		–	
Cotton, wool	2	1096	–		–		–		–		–		–		–		–	
Cristobalite, silicon dioxide	–		–		–		–		–		–		–		–		13	353
Crotonylene, 2-butyne	–		6s	12	–		–		–		–		–		–		–	
Cryolite, halide mineral	–		–		–		8	1660	–		–		–		–		–	
Crystex, sulfur	–		–		–		8	117	–		–		–		–		–	
Crystolon, SiC	–		–		8	800	8	805	–		–		–		–		–	
Cumene	–		6s	22	–		–		–		–		–		–		–	
Cumol	–		6s	22	–		–		–		–		–		–		–	
Cupric oxide	–		–		–		8	247	–		8	249	–		–		–	
Cuprous oxide	–		–		8	243 245	–		–		8	249	–		–		–	
Cuprum	1	68	4	51	–		–		–		–		–		–		–	
Cyanogen	–		6s	24	–		–		–		–		–		–		–	
Cyanogen chloride	–		6s	24	–		–		–		–		–		–		–	
1,4-Cyclohexadienedione	–		6s	2	–		–		–		–		–		–		–	
Cyclohexane	–		6s	25	–		–		–		–		–		–		–	
Cyclohexane-n-hexane, mixture	–		–		–		–		–		–		–		11	408	–	
Cyclohexene	–		6s	25	–		–		–		–		–		–		–	
Cyclohexylmethane	–		6s	62	–		–		–		–		–		–		–	
Cyclopropane	–		6s	26	–		–		–		–		–		–		–	
Cyclopropane-helium, mixture	3	325	–		–		–		–		–		–		–		–	
p-Cymene	–		6s	26	–		–		–		–		–		–		–	
p-Cymol	–		6s	26	–		–		–		–		–		–		–	
n-Decane	3	164	6	170	–		–		–		–		–		–		–	
n-Decane-methane, mixture	–		–		–		–		–		–		–		11	410	–	
Delrin acetal DA-500	–		–		–		–		–		–		–		–		13	1479
Deuteriomethane	–		6s	58	–		–		–		–		–		–		–	
Deuteriotritritiomethane	–		6s	58	–		–		–		–		–		–		–	
Deuterium, D_2	3	21	6	15	–		–		–		–		–		11	13	–	
Deuterium, monatomic	–		6s	26	–		–		–		–		–		–		–	
Deuterium fluoride, DF	–		6s	47	–		–		–		–		–		–		–	
Deuterium hydride, HD	–		6s	48	–		–		–		–		–		–		–	
Deuterium-helium, mixture	3	327	–		–		–		–		–		–		–		–	
Deuterium-hydrogen, mixture	3	407	–		–		–		–		–		–		11	413	–	
Deuterium-hydrogen deuteride, mixture	–		–		–		–		–		–		–		11	415	–	
Deuterium hydrogen sulfide	–		6s	51	–		–		–		–		–		–		–	
Deuterium-krypton, mixture	3	349	–		–		–		–		–		–		–		–	
Deuterium-krypton-neon, mixture	3	491	–		–		–		–		–		–		–		–	
Deuterium-neon, mixture	3	360	–		–		–		–		–		–		–		–	
Deuterium-nitrogen, mixture	3	410	–		–		–		–		–		–		–		–	

Substance Name	Thermal Conductivity		Specif. Heat		Thermal Radiative Properties								Thermal Diffusivity		Viscosity		Thermal Expansion	
					Emissivity		Reflectivity		Absorptivity		Transmissiv.							
	V.	Page	V.	Page	V.	Page	V.	Page	V.	Page	V.	Page	V.	Page	V.	Page	V.	Page
Deuterium-xenon, mixture	3	371	–		–		–		–		–		–		–		–	
Deuterium oxide, D_2O		–	6s	95	–		–		–		–		–		–		13	224
Deuterium selenide		–	6s	49	–		–		–		–		–		–		–	
Deuterium sulfide		–	6s	50	–		–		–		–		–		–		–	
Deuterium tritium sulfide		–	6s	52	–		–		–		–		–		–		–	
Diamine		–	6s	44	–		–		–		–		–		–		–	
1,2-Diaminoethane		–	6s	37	–		–		–		–		–		–		–	
Diamond	2	9	5	4	–		–		–		–		10	24	–		13	19
Diamond, type I	2	10	–		–		–		–		–		–		–		–	
Diamond, type II	2	10	–		–		–		–		–		–		–		–	
Diaspore		–		–	–		–		–		8	1664	–		–		–	
Diatomaceous earth	2	814			–		–		–		–		–		–		–	
Diatomite	2	814			–		–		–		–		–		–		–	
Diatomite aggregate Sil-O-Cel, coarse grade	2	1112			–		–		–		–		–		–		–	
Dibromomethane		–	6s	26	–		–		–		–		–		–		–	
1,2-Dibromoethane		–	6s	26	–		–		–		–		–		–		–	
1,2-Dibromoethane, EDB		–	6s	26	–		–		–		–		–		–		–	
1,2-Dibromopropane		–	6s	27	–		–		–		–		–		–		–	
1,3-Dibromopropane		–	6s	27	–		–		–		–		–		–		–	
Dibutyl ether		–	6s	11	–		–		–		–		–		–		–	
α-Dichloroethane		–	6s	27	–		–		–		–		–		–		–	
sym-Dichloroethane		–	6s	27	–		–		–		–		–		–		–	
1,1-Dichloroethane		–	6s	27	–		–		–		–		–		–		–	
1,2-Dichloroethane		–	6s	27	–		–		–		–		–		–		–	
sym-Dichloroethylene		–	6s	28	–		–		–		–		–		–		–	
1,2-Dichloroethylene		–	6s	28	–		–		–		–		–		–		–	
Dichlorofluoromethane	3	193	6	212	–		–		–		–		–		11	155	–	
1,1-Dichloro-1-fluoroethane		–	6s	28	–		–		–		–		–		–		–	
Dichlorodifluoromethane	3	187	6	204	–		–		–		–		–		11	150	–	
Dichlorofluoromethane, monodeuterated		–	6s	28	–		–		–		–		–		–		–	
Dichloromethane		–	6s	28	–		–		–		–		–		–		–	
1,2-Dichloropropane		–	6s	29	–		–		–		–		–		–		–	
Dichlorotetrafluoroethane		–		–	–		–		–		–		–		11	160	–	
unsym-Dichlorotetrafluoroethane		–	6s	29	–		–		–		–		–		–		–	
1,1-Dichlorotetrafluoroethane		–	6s	29	–		–		–		–		–		–		–	
1,2-Dichloro-1,1,2,2-tetrafluoroethane	3	205	6	228	–		–		–		–		–		–		–	
2,2-Dichloro-1,1,1-trifluoroethane		–	6s	29	–		–		–		–		–		–		–	
Dicopper sulfide + iron sulfide + trinickel disulfide, mixture	2	700	–		–		–		–		–		–		–		–	
Dicopper sulfide + trinickel disulfide, mixture	2	701	–		–		–		–		–		–		–		–	
Dideuterioditritiomethane		–	6s	58	–		–		–		–		–		–		–	
Dideuteriomethane		–	6s	58	–		–		–		–		–		–		–	

Substance Name	Thermal Conductivity V.	Page	Specif. Heat V.	Page	Emissivity V.	Page	Reflectivity V.	Page	Absorptivity V.	Page	Transmissiv. V.	Page	Thermal Diffusivity V.	Page	Viscosity V.	Page	Thermal Expansion V.	Page
Diethylamine-ethyl ether	3	472		–		–		–		–		–		–		–		–
Diethyl carbinol		–	6s	72		–		–		–		–		–		–		–
Diethyl ethanedioate		–	6s	30		–		–		–		–		–		–		–
Diethyl ketone		–	6s	72		–		–		–		–		–		–		–
Diethyl oxalate		–	6s	30		–		–		–		–		–		–		–
Diethylene		–	6s	5		–		–		–		–		–		–		–
Diethylisopropylmethane		–	6s	38		–		–		–		–		–		–		–
Diethylmethylmethane		–	6s	65		–		–		–		–		–		–		–
Diethylpropylmethane		–	6s	38		–		–		–		–		–		–		–
1,1-Difluoroethane		–		–		–		–		–		–		–	11	165		–
1,1-Difluoroethylene		–	6s	30		–		–		–		–		–		–		–
Difluoromethane		–	6s	30		–		–		–		–		–		–		–
Dihexyl		–	6s	34		–		–		–		–		–		–		–
m-Dihydroxybenzene		–	6s	83		–		–		–		–		–		–		–
o-Dihydroxybenzene		–	6s	83		–		–		–		–		–		–		–
p-Dihydroxybenzene		–	6s	53		–		–		–		–		–		–		–
1,2-Dihydroxybenzene		–	6s	83		–		–		–		–		–		–		–
1,3-Dihydroxybenzene		–	6s	83		–		–		–		–		–		–		–
1,4-Dihydroxybenzene		–	6s	53		–		–		–		–		–		–		–
β,β'-Dihydroxydipropyl ether		–	6s	34		–		–		–		–		–		–		–
2,2'-Dihydroxydipropyl ether		–	6s	34		–		–		–		–		–		–		–
1,2-Dihydroxypropane		–	6s	76		–		–		–		–		–		–		–
Diiodomethane		–	6s	30		–		–		–		–		–		–		–
Diisoamyl		–	6s	33		–		–		–		–		–		–		–
Diisobutyl		–	6s	33		–		–		–		–		–		–		–
Diisobutylene		–	6s	94		–		–		–		–		–		–		–
Diisopropyl		–	6s	32		–		–		–		–		–		–		–
Dilver O		–		–		–		–		–		–		–		–	12	710
Dimethyl		–	6	174		–		–		–		–		–		–		–
Dimethyl carbinol		–	6s	79		–		–		–		–		–		–		–
Dimethyl ether		–	6s	63		–		–		–		–		–		–		–
Dimethyl ether-methyl chloride, mixture		–		–		–		–		–		–		–	11	547		–
Dimethyl ether-methyl chloride-sulphur dioxide, mixture		–		–		–		–		–		–		–	11	592		–
Dimethyl ether-sulfur dioxide, mixture		–		–		–		–		–		–		–	11	549		–
Dimethyl ethyl carbinol		–	6s	61		–		–		–		–		–		–		–
Dimethyl ketone	3	129	6	113		–		–		–		–		–		–		–
Dimethyl sulfide		–	6s	69		–		–		–		–		–		–		–
2,5-Dimethyl thiophene		–	6s	33		–		–		–		–		–		–		–
sym-Dimethylacetone		–	6s	72		–		–		–		–		–		–		–
Dimethylacetylene		–	6s	12		–		–		–		–		–		–		–
Dimethylamine		–	6s	30		–		–		–		–		–		–		–
Dimethylamine, DMA		–	6s	30		–		–		–		–		–		–		–

Substance Name	Thermal Conductivity		Specif. Heat		Thermal Radiative Properties								Thermal Diffusivity		Viscosity		Thermal Expansion	
					Emissivity		Reflectivity		Absorptivity		Transmissiv.							
	V.	Page	V.	Page	V.	Page	V.	Page	V.	Page	V.	Page	V.	Page	V.	Page	V.	Page
1,2-Dimethylbenzene		–	6s	99	–		–		–		–		–		–		–	
1,3-Dimethylbenzene		–	6s	98	–		–		–		–		–		–		–	
1,4-Dimethylbenzene		–	6s	101	–		–		–		–		–		–		–	
2,2-Dimethylbutane		–	6s	31	–		–		–		–		–		–		–	
2,3-Dimethylbutane		–	6s	32	–		–		–		–		–		–		–	
1,2-Dimethylcyclopentane		–	6s	33	–		–		–		–		–		–		–	
Dimethylenemethane		–	6s	75	–		–		–		–		–		–		–	
cis-sym-Dimethylethylene		–	6s	9	–		–		–		–		–		–		–	
sym-Dimethylethylene		–	6s	9	–		–		–		–		–		–		–	
trans-sym-Dimethylethylene		–	6s	10	–		–		–		–		–		–		–	
unsym-Dimethylethylene		–	6s	67	–		–		–		–		–		–		–	
2,3-Dimethylhexane		–	6s	33	–		–		–		–		–		–		–	
2,5-Dimethylhexane		–	6s	33	–		–		–		–		–		–		–	
3,3-Dimethylhexane		–	6s	33	–		–		–		–		–		–		–	
3,4-Dimethylhexane		–	6s	33	–		–		–		–		–		–		–	
Dimethylmethane	3	240	6	279	–		–		–		–		–		–		–	
2,7-Dimethyloctane		–	6s	33	–		–		–		–		–		–		–	
Dimethylpropane		–	6s	33	–		–		–		–		–		–		–	
Dimethylpropylmethane		–	6s	64	–		–		–		–		–		–		–	
Diniobium pentoxide + uranium dioxide, mixture	2	427		–	–		–		–		–		–		–		–	
m-Dinitrobenzene		–	6s	34	–		–		–		–		–		–		–	
o-Dinitrobenzene		–	6s	34	–		–		–		–		–		–		–	
p-Dinitrobenzene		–	6s	34	–		–		–		–		–		–		–	
1,2-Dinitrobenzene		–	6s	34	–		–		–		–		–		–		–	
1,3-Dinitrobenzene		–	6s	34	–		–		–		–		–		–		–	
1,4-Dinitrobenzene		–	6s	34	–		–		–		–		–		–		–	
1,2-Dinitrobenzene		–	6s	34	–		–		–		–		–		–		–	
1,3-Dinitrobenzol		–	6s	34	–		–		–		–		–		–		–	
1,4-Dinitrobenzol		–	6s	34	–		–		–		–		–		–		–	
Dioform		–	6s	28	–		–		–		–		–		–		–	
Diopside		–		–	–		–		–		–		–		–		13	708
Dioxane-benzyl acetate, mixture		–		–	–		–		–		–		–		11	512	–	
Diphenyl	2	989		–	–		–		–		–		–		–		–	
Diphenyl oxide	2	990	6s	73	–		–		–		–		–		–		–	
Diphenylamine	2	991		–	–		–		–		–		–		–		–	
unsym-Diphenylethane		–	6s	34	–		–		–		–		–		–		–	
1,1-Diphenylethane		–	6s	34	–		–		–		–		–		–		–	
Diphenylmethane		–	6s	34	–		–		–		–		–		–		–	
Diphenylmethane + naphthalene, mixture	2	994		–	–		–		–		–		–		–		–	
Dipropylene glycol		–	6s	34	–		–		–		–		–		–		–	
Dipropyl ether		–	6s	81	–		–		–		–		–		–		–	
Ditan		–	6s	34	–		–		–		–		–		–		–	

Substance Name	Thermal Conductivity		Specif. Heat		Thermal Radiative Properties										Thermal Diffusivity		Viscosity		Thermal Expansion	
					Emissivity		Reflectivity		Absorptivity		Transmissiv.									
	V.	Page	V.	Page	V.	Page	V.	Page	V.	Page	V.	Page	V.	Page	V.	Page	V.	Page
Ditantalum hydride, Ta_2H		–	5	1040	–		–		–		–		–		–		–	
Ditritiomethane		–	6s	58	–		–		–		–		–		–		–	
Divinyl		–	6s	5	–		–		–		–		–		–		–	
Dodecane		–	6s	34	–		–		–		–		–		–		–	
Dolomite, NTS	2	811		–	–		–		–		–		–		–		–	
Dolomite, NTS No. 1	2	810		–	–		–		–		–		–		–		–	
Dolomite, NTS No. 2	2	811		–	–		–		–		–		–		–		–	
Dow metal	1	999		–	–		–		–		–		–		–		–	
Dracylic acid		–	6s	2	–		–		–		–		–		–		–	
Drakenfeld		–		–	–		–		–		–		–		–		13	554
Dutch liquid		–	6s	27	–		–		–		–		–		–		–	
Dutch oil		–	6s	27	–		–		–		–		–		–		–	
Dysprosia		–	5	83	–		–		–		–		–		–		–	
Dysprosium, Dy	1	82	4	62	–		–		–		–		10	62	–		12	92
Dysprosium alloys:																		
Dy + Ta		–		–	–		–		–		–		–		–		12	804
Dy + Ta + ΣXi		–		–	–		–		–		–		–		–		12	1120
Dysprosium aluminum garnet		–		–	–		–		–		–		–		–		13	476
Dysprosium aluminum oxides:																		
$Dy_2O_3 \cdot 2Al_2O_3$		–		–	–		–		–		–		–		–		13	475
$3Dy_2O_3 \cdot 5Al_2O_3$		–		–	–		–		–		–		–		–		13	476
Dysprosium borides:																		
DyB_6		–		–	–		–		–		–		8	727	–		–	
DyB_{12}		–		–	–		–		–		–		–		–		13	793
Dysprosium carbide, DyC_2		–		–	–		–		–		–		–		–		13	935
Dysprosium chloride hexahydrate, $DyCl_3 \cdot 6H_2O$		–	5	818	–		–		–		–		–		–		–	
Dysprosium niobium oxide, $Dy_2O_3 \cdot Nb_2O_5$		–		–	–		–		–		–		–		–		13	525
Dysprosium nitride, DyN		–		–	–		–		–		–		8	1090	–		–	
Dysprosium oxide, Dy_2O_3		–	5	83	–		8	252	8	254	–		–		–		13	227
Earth:	2	813		–	–		–		–		–		–		–		–	
Diatomaceous	2	814		–	–		–		–		–		–		–		–	
Kieselguhr	2	814		–	–		–		–		–		–		–		–	
Kieselguhr, ignited	2	814		–	–		–		–		–		–		–		–	
Kieselguhr, ordinary	2	814		–	–		–		–		–		–		–		–	
Ebonite rubber	2	971		–	–		–		–		–		10	617	–		–	
Egg		–		–	–		–		–		–		10	633	–		–	
Eggplant		–		–	–		–		–		–		10	634	–		–	
Elastomer rubber	2	974		–	–		–		–		–		–		–		–	
Enamel, ACME no. 800		–		–	–		9	483	–		–		–		–		–	
Enamel, ACME no. 801		–		–	–		9	483	–		–		–		–		–	
Enamel, ACME no. 803		–		–	–		9	483	–		–		–		–		–	

Substance Name	Thermal Conduc-tivity		Specif. Heat		Thermal Radiative Properties								Thermal Diffu-sivity		Visco-sity		Thermal Expan-sion	
					Emis-sivity		Reflec-tivity		Absorp-tivity		Trans-missiv.							
	V.	Page	V.	Page	V.	Page	V.	Page	V.	Page	V.	Page	V.	Page	V.	Page	V.	Page
Enamel, cerium dioxide + magnesium oxide pigment in NBS Frit no. 332 binder	–		–		–		9	452	–		–		–		–		–	
Enamel, cerium dioxide + cobalt oxide pigment in base glaze No. 3 binder	–		–		9	445	–		–		–		–		–		–	
Enamel, cerium dioxide pigment in:																		
Barium beryllium silicate binder	–		–		9	445 448	–		–		–		–		–		–	
Base glaze no. 1 binder	–		–		9	445	–		–		–		–		–		–	
Base glaze no. 3 binder	–		–		9	445	–		–		–		–		–		–	
Enamel, cerium dioxide + tin oxide pigment in NBS Frit no. 332 binder	–		–		9	445	9	452	–		–		–		–		–	
Enamel, cerium dioxide + zirconium dioxide pigment in NBS Frit no. 332 binder	–		–		–		9	452	–		–		–		–		–	
Enamel, chromium oxide + cobalt oxide pigment in base glaze No. 1 binder	–		–		9	455	–		–		–		–		–		–	
Enamel, chromium oxide + iron oxide pigment in base glaze No. 3 binder	–		–		9	456	–		–		–		–		–		–	
Enamel, chromium oxide pigment in:																		
Barium borosilicate frit binder	–		–		9	455 459	–		–		–		–		–		–	
Base glaze No. 1 binder	–		-		9	455	–		–		–		–		–		–	
Base glaze No. 2 binder	–		–		9	455	–		–		–		–		–		–	
Base glaze No. 3 binder	–		–		9	455	–		–		–		–		–		–	
NBS frit No. 332 binder	–		–		9	455	9	463	–		–		–		–		–	
Enamel, cobalt oxide + chromium oxide pigment in base glaze No. 1 binder	–		–		9	464	–		–		–		–		–		–	
Enamel, cobalt oxide + chromium oxide pigment in base glaze No. 2 binder	–		–		9	464	–		–		–		–		–		–	
Enamel, cobalt oxide + manganese oxide pigment in base glaze No. 2 binder	–		–		9	464	–		–		–		–		–		–	
Enamel, cobalt oxide + nickel oxide pigment in base glaze No. 3 binder	–		–		9	464	–		–		–		–		–		–	
Enamel, cobalt oxide pigment in base glaze No. 3 binder	–		–		9	464	–		–		–		–		–		–	
Enamel, $CoO \cdot Cr_2O_3$ spinel pigment in NBS Frit No. 332 binder	–		–		9	472	9	475	–		–		–		–		–	
Enamel, $CoO \cdot Fe_2O_3$ spinel pigment in NBS Frit No. 332 binder	–		–		9	472	9	475	–		–		–		–		–	
Enamel, $CoO \cdot Mn_2O_3$ spinel pigment in NBS Frit No. 332 binder	–		–		9	472	9	475	–		–		–		–		–	
Enamel, DaCote black	–		–		–		9	542	–		–		–		–		–	
Enamel, dreem No. 13 N27ES4	–		–		–		9	495	–		–		–		–		–	
Enamel, ferro white porcelain	–		–		9	576	–		–		–		–		–		–	
Enamel, iron oxide + chromium oxide pigment in base glaze No. 3 binder	–		–		9	466	–		–		–		–		–		–	

| Substance Name | Thermal Conductivity | | Specif. Heat | | Thermal Radiative Properties | | | | | | | | Thermal Diffusivity | | Viscosity | | Thermal Expansion | |
| | | | | | Emissivity | | Reflectivity | | Absorptivity | | Transmissiv. | | | | | | | |
	V.	Page	V.	Page	V.	Page	V.	Page	V.	Page	V.	Page	V.	Page	V.	Page	V.	Page
Enamel, iron oxide + cobalt oxide + chromium oxide pigment in NBS Frit No. 332 binder		–		–	9	466	9	467		–		–		–		–		–
Enamel, iron oxide + nickel oxide pigment in base glaze No. 3 binder		–		–	9	466		–		–		–		–		–		–
Enamel, Magic Iron Cement Co. white porcelain		–		–		–	9	528		–		–		–		–		–
Enamel, manganese oxide + cobalt oxide pigment in base glaze No. 2 binder		–		–	9	468		–		–		–		–		–		–
Enamel, manganese oxide + iron oxide pigment in base glaze No. 3 binder		–		–	9	468		–		–		–		–		–		–
Enamel, metallic kerpo QD No. WB-S-N 52-E-4		–		–		–	9	520		–		–		–		–		–
Enamel, no. 102, broma metallic		–		–		–	9	489		–		–		–		–		–
Enamel, no. 113, broma alkyd		–		–		–	9	488		–		–		–		–		–
Enamel, P-110 white porcelain		–		–		–	9	578		–		–		–		–		–
Enamel, Pittsburgh flat white undercoater, LA-404		–		–		–	9	552		–		–		–		–		–
Enamel, potassium titanate porcelain		–		–	9	470		–		–		–		–		–		–
Enamel, silicon, Heyden L28	2	921		–		–		–		–		–		–		–		–
Enstatite		–		–		–	8	1689	8	1692		–		–		–	13	717
Epon 815		–		–		–		–		–		–		–		–	13	1504
Epon 828		–		–		–		–		–		–		–		–	13	1502
Epoxy resin, Marglas 555		–		–		–		–		–		–		–		–	13	1505
Epoxy resin, Standard 43		–		–		–		–		–		–		–		–	13	1504
1,2-Epoxyethane		–	6s	37		–		–		–		–		–		–		–
Erbia		–	5	86		–		–		–		–		–		–		–
Erbium, Er	1	86	4	65	7	202		–		–		–	10	65		–	12	98
Erbium borides:																		
Er_6		–		–		–		–		–	8	727		–		–		–
ErB_{12}		–		–		–		–		–		–		–		–	13	755
Erbium carbide, ErC_2		–		–		–		–		–		–		–		–	13	935
Erbium chloride hexahydrate, $ErCl_3 \cdot 6H_2O$		–	5	822		–		–		–		–		–		–		–
Erbium gallate		–	5	1440		–		–		–		–		–		–		–
Erbium gallium oxide, $Er_3Ga_5O_{12}$, garnet		–	5	1440		–		–		–		–		–		–		–
Erbium nitride, ErN		–		–		–		–		–	8	1090		–		–		–
Erbium oxide, Er_2O_3		–	5	86	8	255 257 259 261		–		–		–		–		–	13	231
Erythrene		–	6s	5		–		–		–		–		–		–		–
Ethanal		–	6s	1		–		–		–		–		–		–		–
Ethane	3	167	6	174		–		–		–		–		–	11	167		–
Ethane, hexadeuterated		–	6s	35		–		–		–		–		–		–		–

Substance Name	Thermal Conductivity V.	Page	Specif. Heat V.	Page	Thermal Radiative Properties Emissivity V.	Page	Reflectivity V.	Page	Absorptivity V.	Page	Transmissiv. V.	Page	Thermal Diffusivity V.	Page	Viscosity V.	Page	Thermal Expansion V.	Page
1,2-Ethanediamine		–	6s	37	–		–		–		–		–			–		–
Ethanedinitrile		–	6s	24	–		–		–		–		–			–		–
1,2-Ethanediol		–	6	192	–		–		–		–		–			–		–
Ethane-ethylene, mixture		–		–	–		–		–		–		–		11	417		–
Ethane-helium, mixture	3	329		–	–		–		–		–		–			–		–
Ethane-hydrogen, mixture		–		–	–		–		–		–		–		11	419		–
Ethane-methane, mixture		–		–	–		–		–		–		–		11	421		–
Ethane-methane-nitrogen-propane, mixture		–		–	–		–		–		–		–		11	596		–
Ethane-propane, mixture		–		–	–		–		–		–		–		11	423		–
Ethanethiol		–	6s	35	–		–		–		–		–			–		–
Ethanoic acid		–	6s	1	–		–		–		–		–			–		–
Ethanol *Dimethyl ether*-methyl formate, mixture	3	169		–	–		–		–		–		–			–		–
Ethanol-methyl formate, mixture	3	474		–	–		–		–		–		–			–		–
Dimethyl ether-propane, mixture	3	456		–	–		–		–		–		–			–		–
Ethene		–	6	185	–		–		–		–		–			–		–
Ethenone		–	6s	57	–		–		–		–		–			–		–
Ethenyl ethanoate		–	6s	95	–		–		–		–		–			–		–
Ether, wood		–	6s	63	–		–		–		–		–			–		–
Ethine		–	6	117	–		–		–		–		–			–		–
Ethinyl trichloride		–	6s	91	–		–		–		–		–			–		–
Ethoxyethane		–	6	194	–		–		–		–		–			–		–
Ethyl acetate		–	6s	35	–		–		–		–		–			–		–
Ethyl acetylene		–	6s	11	–		–		–		–		–			–		–
Ethyl alcohol	3	169	6	180	–		–		–		–		–		11	172		–
Ethyl aldehyde		–	6s	1	–		–		–		–		–			–		–
Ethyl benzene		–	6s	35	–		–		–		–		–			–		–
Ethyl bromide		–	6s	4	–		–		–		–		–			–		–
Ethyl butanoate		–	6s	37	–		–		–		–		–			–		–
Ethyl butyrate		–	6s	37	–		–		–		–		–			–		–
Ethyl carbinol		–	6s	77	–		–		–		–		–			–		–
Ethyl chloride		–	6s	21	–		–		–		–		–			–		–
Ethyl dimethylmethane		–	6s	59	–		–		–		–		–			–		–
Ethyl ether	3	179		–	–		–		–		–		–			–		–
Ethyl ether-hydrogen, mixture		–		–	–		–		–		–		–		11	519		–
Ethyl ethylene		–	6s	8	–		–		–		–		–			–		–
Ethyl fluoride		–	6s	41	–		–		–		–		–			–		–
Ethyl formate		–	6s	38	–		–		–		–		–			–		–
3-Ethyl hexane		–	6s	38	–		–		–		–		–			–		–
Ethyl hydrosulfide		–	6s	35	–		–		–		–		–			–		–
Ethyl isobutyl methane		–	6s	64	–		–		–		–		–			–		–
Ethyl isovalerate		–	6s	38	–		–		–		–		–			–		–
Ethyl ketone		–	6s	72	–		–		–		–		–			–		–

| Substance Name | Thermal Conductivity | | Specif. Heat | | Thermal Radiative Properties | | | | | | | | Thermal Diffusivity | | Viscosity | | Thermal Expansion | |
| | | | | | Emissivity | | Reflectivity | | Absorptivity | | Transmissiv. | | | | | | | |
	V.	Page	V.	Page	V.	Page	V.	Page	V.	Page	V.	Page	V.	Page	V.	Page	V.	Page
Ethyl mercaptan		–	6s	35		–		–		–		–		–		–		–
Ethyl methanoate		–	6s	38		–		–		–		–		–		–		–
Ethyl methyl acetylene		–	6s	73		–		–		–		–		–		–		–
Ethyl methyl carbinol		–	6s	7		–		–		–		–		–		–		–
Ethyl methyl ketone		–	6s	7		–		–		–		–		–		–		–
3-Ethyl-2-methylpentane		–	6s	38		–		–		–		–		–		–		–
3-Ethyl-3-methylpentane		–	6s	38		–		–		–		–		–		–		–
Ethyl oxalate		–	6s	30		–		–		–		–		–		–		–
Ethyl oxide		–	6	194		–		–		–		–		–		–		–
Ethyl propanoate		–	6s	38		–		–		–		–		–		–		–
1-Ethyl-1-propanol		–	6s	72		–		–		–		–		–		–		–
Ethyl propionate		–	6s	38		–		–		–		–		–		–		–
Ethyl sulfhydrate		–	6s	35		–		–		–		–		–		–		–
Ethyl thioalcohol		–	6s	35		–		–		–		–		–		–		–
Ethyl trimethylmethane		–	6s	31		–		–		–		–		–		–		–
Ethyl valerianate		–	6s	38		–		–		–		–		–		–		–
Ethylene	3	173	6	185		–		–		–		–		–	11	174		
Ethylene alcohol		–	6	192		–		–		–		–		–		–		–
Ethylene bromide		–	6s	26		–		–		–		–		–		–		–
Ethylene dibromide		–	6s	26		–		–		–		–		–		–		–
Ethylene glycol	3	177	6	192		–		–		–		–		–		–		–
Ethylene-helium, mixture	3	331		–		–		–		–		–		–		–		–
Ethylene-hydrogen, mixture	3	413		–		–		–		–		–		–	11	425		–
Ethylene-methane, mixture	3	415		–		–		–		–		–		–	11	428		–
Ethylene-nitrogen, mixture	3	417		–		–		–		–		–		–	11	432		–
Ethylene oxide		–	6s	37		–		–		–		–		–		–		–
Ethylene-oxygen, mixture		–		–		–		–		–		–		–	11	434		–
Ethylenediamine		–	6s	37		–		–		–		–		–		–		–
Ethyl ethanoate		–	6s / 6	35 / 194		–		–		–		–		–	11	180		–
Ethyne		–	6	117		–		–		–		–		–		–		–
Eureka	1	563		–		–		–		–		–		–		–		–
Europium, Eu	1	90	4	68		–		–		–		–	10	66		–	12	104
Europium boride, EuB$_6$		–		–	8	723		–		–		–		–		–		–
Europium oxides:																		
EuO		–		–		–		–		–		–		–		–	13	235
Eu$_2$O$_3$		–	5	89		–	8	546		–		–		–		–	13	236
Europium selenide, EuSe		–		–		–		–		–		–		–		–	13	1192
Europium silicate		–		–		–		–		–	8	622		–		–		–
Euopium sulfate octahydrate, Eu$_2$(SO$_4$)$_3 \cdot$8H$_2$O		–	5	1197		–		–		–		–		–		–		–
Europium sulfide, EuS		–		–		–	8	1234		–		–		–		–	13	1239
Excelsior	2	1113		–		–		–		–		–		–		–		–

Substance Name	Thermal Conductivity		Specif. Heat		Thermal Radiative Properties								Thermal Diffusivity		Viscosity		Thermal Expansion	
					Emissivity		Reflectivity		Absorptivity		Transmissiv.							
	V.	Page	V.	Page	V.	Page	V.	Page	V.	Page	V.	Page	V.	Page	V.	Page	V.	Page
Fabric, rené 41 cloth	2	1102	–		–		–		–		–		–		–		–	
Fat, beef	2	1072	–		–		–		–		–		–		–		–	
Fat, bone	2	1072	–		–		–		–		–		–		–		–	
Fat, pig	2	1073	–		–		–		–		–		–		–		–	
Fayalite	–		–		–		8	1689	–		–		–		–		13	710
Ferric oxide	–		5	110	8	280 282	8	284	–		–		10	391	–		–	
Ferrosoferric oxide	–		5	114	8	282	–		–		–		–		–		–	
Ferrous alloys	–		–		–		–		–		–		10	331	–		–	
Ferrous chloride	–		5	832	–		–		–		–		–		–		–	
Ferrous fluoride	–		5	940	–		–		–		–		–		–		–	
Ferrous oxide	–		5	107	–		–		–		–		–		–		–	
Fiber, mineral	2	1139	–		–		–		–		–		–		–		–	
Fiber, redwood	2	1091	–		–		–		–		–		–		–		–	
Fiber, vulcanized	2	1088	–		–		–		–		–		–		–		–	
Fiberboard, cellulose	2	1110	–		–		–		–		–		–		–		–	
Fiberboard, vegetable	2	1129	–		–		–		–		–		–		–		–	
Fiberglass, blanket	2	1115	–		–		–		–		–		–		–		–	
Fiberglass, insulation	2	1117	–		–		–		–		–		–		–		–	
Fiberglass laminates, epoxy reinforced	–		–		–		–		–		–		10	559	–		–	
Fiberite	2	1052	–		–		–		–		–		–		–		–	
Fibers, wood	2	1091	–		–		–		–		–		–		–		–	
Fir	2	1073	–		–		–		–		–		–		–		–	
Fir, plywood	2	1114	–		–		–		–		–		–		–		–	
Firebrick, Missouri	2	905	–		–		–		–		–		–		–		–	
Fish	–		–		–		–		–		–		10	635	–		–	
Fluon, poly(tetrafluoroethylene)	–		–		–		–		–		–		–		–		13	1446
Fluophosgene	–		6s	16	–		–		–		–		–		–		–	
Fluorethane	–		6s	41	–		–		–		–		–		–		–	
Fluorides, cubic perovskite, miscellaneous	–		–		–		8	987	–		8	989	–		–		–	
Fluorides, miscellaneous	–		–		–		8	991	–		8	993	–		–		–	
Fluorine, F_2	3	26	6	19	–		–		–		–		–		11	16	–	
Fluorine, monatomic	–		6s	38	–		–		–		–		–		–		–	
Fluorite	–		–		–		–		–		8	933	–		–		13	1027
Fluorobenzene	–		6s	41	–		–		–		–		–		–		–	
Fluorocarbons	–		–		–		–		–		–		10	588	–		–	
Fluoroethylene	–		6s	41	–		–		–		–		–		–		–	
Fluoroform, monodeuterated	–		6s	42	–		–		–		–		–		–		–	
Fluoroformyl fluoride	–		6s	16	–		–		–		–		–		–		–	
Fluoromethane	–		6s	42	–		–		–		–		–		–		–	
Fluoroplast IV	–		–		–		–		–		–		–		–		13	1446
Fluorothene	–		–		–		–		–		–		–		–		13	1442
Fluorotribromomethane	–		6s	91	–		–		–		–		–		–		–	

Substance Name	Thermal Conductivity		Specif. Heat		Thermal Radiative Properties								Thermal Diffusivity		Viscosity		Thermal Expansion	
					Emissivity		Reflectivity		Absorptivity		Transmissiv.							
	V.	Page	V.	Page	V.	Page	V.	Page	V.	Page	V.	Page	V.	Page	V.	Page	V.	Page
Formaldehyde		–	6s	42		–		–		–		–		–		–		–
Formalin		–	6s	42		–		–		–		–		–		–		–
Formalith		–	6s	42		–		–		–		–		–		–		–
Formic aldehyde		–	6s	42		–		–		–		–		–		–		–
Formic ether		–	6s	38		–		–		–		–		–		–		–
Formol		–	6s	42		–		–		–		–		–		–		–
Formonitrile		–	6s	46		–		–		–		–		–		–		–
Formyl		–	6s	43		–		–		–		–		–		–		–
Formyl tribromide		–	6s	5		–		–		–		–		–		–		–
Forsterite, Mg_2SiO_4	2	275		–		–	8	1689	8	1692		–		–		–	13	720
Frenchtown 4402		–		–	8	144 147		–		–		–		–		–		–
Freon 10	3	156	6	159		–		–		–		–		–		–		–
Freon 11	3	183	6	200		–		–		–		–		–		–		–
Freon 12	3	187	6	204		–		–		–		–		–		–		–
Freon 13	3	191	6	210		–		–		–		–		–		–		–
Freon 20	3	161	6	166		–		–		–		–		–		–		–
Freon 21	3	193	6	212		–		–		–		–		–		–		–
Freon 22	3	197	6	218		–		–		–		–		–		–		–
Freon 30		–	6s	28		–		–		–		–		–		–		–
Freon 31		–	6s	21		–		–		–		–		–		–		–
Freon 41		–	6s	42		–		–		–		–		–		–		–
Freon 112		–	6s	90		–		–		–		–		–		–		–
Freon 113	3	201	6	224		–		–		–		–		–		–		–
Freon 114	3	205	6	228		–		–		–		–		–		–		–
Freon 116		–	6s	44		–		–		–		–		–		–		–
Freon 123		–	6s	29		–		–		–		–		–		–		–
Freon 130		–	6s	90		–		–		–		–		–		–		–
Freon 140		–	6s	91		–		–		–		–		–		–		–
Freon 141		–	6s	28		–		–		–		–		–		–		–
Freon 143		–	6s	92		–		–		–		–		–		–		–
Freon 161		–	6s	41		–		–		–		–		–		–		–
Frigen 114		–	6s	29		–		–		–		–		–		–		–
Fuel, SNAP		–		–		–		–		–		–	10	541		–		–
Furan		–	6s	43		–		–		–		–		–		–		–
2-Furan carbinol		–	6s	43		–		–		–		–		–		–		–
Furfuralcohol		–	6s	43		–		–		–		–		–		–		–
Furfuran		–	6s	43		–		–		–		–		–		–		–
Furfuryl alcohol		–	6s	43		–		–		–		–		–		–		–
α-Furfuryl carbinol		–	6s	43		–		–		–		–		–		–		–
Gabbro	2	816		–		–	8	1681		–		–		–		–		–
Gadolinium, Gd	1	93	4	72		–	7	204		–	7	207	10	67		–	12	107
Gadolinium boride, GdB_6		–		–	8	723		–		–	8	727		–		–		–

Substance Name	Thermal Conductivity		Specif. Heat		Thermal Radiative Properties								Thermal Diffusivity		Viscosity		Thermal Expansion	
					Emissivity		Reflectivity		Absorptivity		Transmissiv.							
	V.	Page	V.	Page	V.	Page	V.	Page	V.	Page	V.	Page	V.	Page	V.	Page	V.	Page
Gadolinium carbide, GdC_2	–		–		–		–		–		–		–		–		13	935
Gadolinium chloride hexahydrate, $GdCl_3 \cdot 6H_2O$	–		5	826	–		–		–		–		–		–		–	
Gadolinium-indium intermetallic compound, $GdIn_3$	–		–		–		–		–		–		–		–		12	526 529 530
Gadolinium-iron intermetallic compound, Gd_2Fe_{17}	–		–		–		–		–		–		–		–		12	527 529 531
Gadolinium molybdenum oxide, $Gd_2O_3 \cdot 3MoO_3$	–		–		–		–		–		–		–		–		13	520
Gadolinium nitrate hexahydrate, $Gd(NO_3)_3 \cdot 6H_2O$	–		5	1142	–		–		–		–		–		–		–	
Gadolinium oxide, Gd_2O_3	–		5	92	8	263 265 267 269	–		–		–		–		–		13	239
Gadolinium-palladium intermetallic compound, $GdPd_3$	–		–		–		–		–		–		–		–		12	528 529 532
Gadolinium oxide + samarium oxide, mixture	2	356	–		–		–		–		–		–		–		–	
Gadolinium oxide + terbium, cermet	–		–		–		–		–		–		–		–		13	1341
Gadolinium silicide, $GdSi_2$	–		–		8	1173	–		–		–		..		–		–	
Galena	–		–		–		8	1199	–		–		–		–		–	
Gallium, Ga	1	97	4	75	–		7	210	7	213	7	216	10	68	–		12	115
Gallium antimonide, GaSb	–		5	300	–		8	1287	–		8	1290	–		–		–	
Gallium arsenate, $GaAsO_4$	–		–		–		–		–		–		–		–		13	619
Gallium arsenic phosphide	–		–		–		8	1107	–		–		–		–		–	
Gallium arsenide, GaAs	1	1277	5	307	–		8	679	–		8	683	10	462	–		13	747
Gallium arsenide + gallium phosphide, mixture	1	1423	–		–		–		–		–		–		–		–	
Gallium-nickel intermetallic compound, GaNi	–		–		–		–		–		–		–		–		12	533
Gallium nitride	–		–		8	1087	–		–		–		–		–		–	
Gallium oxide, Ga_2O_3	–		5	95	–		–		–		–		–		–		–	
Gallium phosphate, $GaPO_4$	–		–		–		–		–		–		–		–		13	690
Gallium phosphide, GaP	–		5	520	8	1092	8	1094	–		8	1098	–		–		13	1168
Gallium selenide, Ga_2Se_3	–		–		–		–		–		–		–		–		13	1192
Gallium-silver intermetallic compound, $GaAg_3$	–		–		–		–		–		–		–		–		12	536
Gallium telluride, Ga_2Te_3	–		5	723	–		–		–		–		–		–		–	
Garnet	2	278	–		–		–		–		–		–		–		–	
Garnet, dysprosium aluminum oxide	–		–		–		–		–		–		–		–		13	476
Garnet, $Er_3Ga_5O_{12}$	–		5	1440	–		–		–		–		–		–		–	
Garnet, yttrium aluminate	–		–		–		8	579	–		–		–		–		–	
Garnet, yttrium aluminate, YAG	–		–		–		8	579	–		–		–		–		–	
Garnet, yttrium ferrate	2	311	–		–		–		–		–		–		–		–	
Gas, laughing	3	114	6	92	–		–		–		–		–		–		–	

Substance Name	Thermal Conductivity		Specif. Heat		Thermal Radiative Properties								Thermal Diffusivity		Viscosity		Thermal Expansion	
					Emissivity		Reflectivity		Absorptivity		Transmissiv.							
	V.	Page	V.	Page	V.	Page	V.	Page	V.	Page	V.	Page	V.	Page	V.	Page	V.	Page
Gehlenite		–		–		–		–		–		–		–		–	13	727
Genetron 11	3	183	6	200		–		–		–		–		–		–		–
Genetron 12	3	187	6	204		–		–		–		–		–		–		–
Genetron 13	3	191	6	210		–		–		–		–		–		–		–
Genetron 22	3	197	6	218		–		–		–		–		–		–		–
Genetron 31		–	6s	21		–		–		–		–		–		–		–
Genetron 113	3	201	6	224		–		–		–		–		–		–		–
Genetron 114	3	205	6	228		–		–		–		–		–		–		–
Genetron 123		–	6s	29		–		–		–		–		–		–		–
Genetron 141		–	6s	28		–		–		–		–		–		~		–
Genetron 1132A		–	6s	30		–		–		–		–		–		–		–
Germanium, Ge	1	108	4	79	7	219 222 224	7	231		–	7	236 240	10	69		–	12	116
Germanium 74, enriched	1	112		–		–		–		–		–		–		–		–
Germanium alloy, Ge + Si	1	597		–		–		–		–		–	10	241		–		–
Germanium hydride, GeH_4		–	5	1033		–		–		–		–		–		–		–
Germanium-lanthanum intermetallic compound, GeLa		–		–		–		–		–		–		–		–	12	537
Germanium-magnesium intermetallic compound, $GeMg_2$		–		–		–		–		–		–		–		–	12	538
Germanium oxides:																		
GeO		–		–		–		–		–	8	549		–		–		–
GeO_2		–	5	98		–	8	271		–		–		–		–		–
Quartz type		–		–		–		–		–		–		–		–	13	243
Rutile type		–		–		–		–		–		–		–		–	13	247
Germanium-praseodymium intermetallic compounds:																		
GePr		–		–		–		–		–		–		–		–	12	540 542 544
Ge_2Pr		–		–		–		–		–		–		–		–	12	539 542 543
Ge_3Pr_5		–		–		–		–		–		–		–		–	12	541 542 545
Germanium silicide, nonstoichiometric		–	5	574		–		–		–		–		–		–		–
Germanium telluride, GeTe	1	1280		–		–	8	1250		–		–		–		–	13	1270
Glass, aluminosilicate		–	5	1227		–		–		–		–		–		–		–
Glass, aluminosilicate 723	2	923		–		–		–		–		–		–		–		–
Glass, aluminum silicate		–		–	8	1523 1525	8	1525	8	1527	8	1528 1530		–		–		–
Glass, amber	2	924		–		–		–		–		–		–		–		–
Glass, AO 1053		–		–		–		–		–	8	1533		–		–		–
Glass, arsenic		–		–		–		–		–	8	1535		–		–		–
Glass, arsenic-selenium		–		–		–		–		–		–		–		–	13	1346
Glass, barium borate		–		–		–		–		–		–		–		–	13	1349

Substance Name	Thermal Conductivity		Specif. Heat		Thermal Radiative Properties								Thermal Diffusivity		Viscosity		Thermal Expansion	
					Emissivity		Reflectivity		Absorptivity		Transmissiv.							
	V.	Page	V.	Page	V.	Page	V.	Page	V.	Page	V.	Page	V.	Page	V.	Page	V.	Page
Glass, boric oxide		–		–		–		–		–		–		–		–	13	1352
Glass, borosilicate	2	923 924	5	1230	8	1539 1541	8	1543	8	1545	8	1546 1547 1549		–		–	13	1355
Glass, borosilicate, 3235	2	923		–		–		–		–		–		–		–		–
Glass, borosilicate, crown	2	923		–		–		–		–		–		–		–		–
Glass, borosilicate, pyrex 7740		–		–		–		–		–		–		–		–	13	1356
Glass, calcium aluminate		–		–		–	8	1551		–	8	1553		–		–		–
Glass, calcium borate		–		–		–		–		–		–		–		–	13	1358
Glass, cellular	2	923		–		–		–		–		–		–		–		–
Glass, ceramics cercor code 9690		–		–		–		–		–		–	10	583		–		–
Glass, colorless	2	924		–		–		–		–		–		–		–		–
Glass, Corning 0080	2	511 928		–		–		–		–		–	10	441 578		–	13	1360
Glass, Corning 0160		–		–		–		–		–	8	1642		–		–		–
Glass, Corning 1173		–		–		–		–		–		–		–		–	13	1361
Glass, Corning 1723		–	5	1227	8	1524	8	1526	8	1527	8	1529 1531	10	432 578		–		–
Glass, Corning 7570		–		–		–		–		–		–		–		–	13	1362
Glass, Corning 7740	2	933		–	8	1580	8	1588	8	1589	8	1590 1594	10	437 578		–		–
Glass, Corning 7900		–		–	8	1622	8	1626	8	1627	8	1629 1633	10	436 578		–		–
Glass, Corning 7905		–		–		–		–		–	8	1633		–		–		–
Glass, Corning 7910		–		–		–		–		–	8	1633		–		–		–
Glass, Corning 7940		–		–	8	1570	8	1572	8	1573	8	1575 1577 1578		–		–		–
Glass, Corning 8325		–		–		–		–		–		–	10	435 578		–		–
Glass, Corning 8362		–		–		–		–		–		–	10	439 578		–		–
Glass, Corning 8363		–		–		–		–		–	8	1642	10	579		–		–
Glass, Corning 8370		–		–		–		–		–		–	10	579		–		–
Glass, Corning 9606		–		–		–		–		–		–		–		–	13	1363
Glass, Corning 9690		–		–	8	1639		–		–		–		–		–		–
Glass, Corning 9752		–		–	8	1637 1638	8	1640		–		–		–		–		–
Glass, Corning 9863		–		–		–		–		–	8	1642		–		–		–
Glass, electroconducting		–		–	8	1559	8	1561	8	1563	8	1564 1566		–		–		–
Glass, flint		–		–	8	1644	8	1646		–	8	1648		–		–		–
Glass, foam	2	924 925		–		–		–		–		–	10	579		–		–
Glass, fused silica	2	925		–	8	1568 1569	8	1571	8	1573	8	1574 1576 1578		–		–		–
Glass, green	2	923		–		–		–		–		–		–		–		–
Glass, high silica		–	5	1234		–		–		–		–		–		–		–
Glass, Jena		–		–		–		–		–		–		–		–	13	1364

Substance Name	Thermal Conductivity		Specif. Heat		Thermal Radiative Properties								Thermal Diffusivity		Viscosity		Thermal Expansion	
					Emissivity		Reflectivity		Absorptivity		Transmissiv.							
	V.	Page	V.	Page	V.	Page	V.	Page	V.	Page	V.	Page	V.	Page	V.	Page	V.	Page
Glass, Jena Gerate	2	924		-		-		-		-		-		-		-		-
Glass, kimble N-51A		-		-	8	1542	8	1544		-	8	1548		-		-		-
Glass, L.O.F. 81E 19778		-		-	8	1560	8	1562	8	1563	8	1565 1567		-		-		-
Glass, L.OF. PB 19195		-		-	8	1560	8	1562	8	1563	8	1565 1567		-		-		-
Glass, lead	2	923		-		-		-		-		-		-		-		-
Glass, lime		-		-		-		-		-		-		-		-	13	1366
Glass, monax	2	924		-		-		-		-		-		-		-		-
Glass, Phoenix	2	924		-		-		-		-		-		-		-		-
Glass, Pittsburgh no. 3235		-	5	1232		-		-		-		-		-		-		-
Glass, plate	2	923 924 925 926	5	1241		-		-		-		-	10	579		-		-
Glass, plate, golden	2	924		-		-		-		-		-		-		-		-
Glass, plate No. 9330, Libbey-Owens-Ford		-	5	1241		-		-		-		-		-		-		-
Glass, pyrex	2	499 923 924 926 927	5	1230		-		-		-		-		-		-	13	1369
Glass, pyrex, ordinary		-		-		-		-		-		-		-		-	13	1371
Glass, pyrex 774	2	923 925	5	1232	8	1582 1584	8	1586 1588		-	8	1592		-		-		-
Glass, pyrex 7740	2	499 924 926		-		-		-		-		-	10	578		-	13	1356 1371
Glass, pyrex G702-EJ		-		-		-		-		-		-		-		-	13	1371
Glass, pyroceram		-	5	1237		-		-		-		-		-		-		-
Glass, Schott BK-8		-		-		-		-		-	8	1533		-		-		-
Glass, Schott B and L 529516		-		-		-		-		-	8	1533		-		-		-
Glass, Schott K5		-		-		-		-		-	8	1533		-		-		-
Glass, Schott KZ SF-4		-		-		-		-		-	8	1533		-		-		-
Glass, Schott SF-2		-		-		-		-		-	8	1533		-		-		-
Glass, silica	2	923 925 926	5	202		-		-		-		-		-		-		-
Glass, silicate	2	511		-		-		-		-		-		-		-		-
Glass, soda	2	923		-		-		-		-		-		-		-		-
Glass, soda lime	2	926	5	1240	8	1609	8	1612	8	1614	8	1615 1617		-		-		-
Glass, soda lime C.G.W. code 0080		-		-		-		-		-		-		-		-	13	1360
Glass, soda lime plate	2	926		-		-		-		-		-		-		-		-
Glass, soda lime silica	2	511 924 927	5	1240		-		-		-		-		-		-		-
Glass, soda lime silica plate	2	923		-		-		-		-		-		-		-		-
Glass, soda silica		-		-		-		-		-	8	1650		-		-		-
Glass, sodium borate		-		-		-		-		-		-		-		-	13	1374

Substance Name	Thermal Conductivity		Specif. Heat		Thermal Radiative Properties								Thermal Diffusivity		Viscosity		Thermal Expansion	
					Emissivity		Reflectivity		Absorptivity		Transmissiv.							
	V.	Page	V.	Page	V.	Page	V.	Page	V.	Page	V.	Page	V.	Page	V.	Page	V.	Page
NaBO$_2$		–	5	1552		–		–		–		–		–		–		–
Na$_2$B$_4$O$_7$		–	5	1556		–		–		–		–		–		–		–
Glass, sodium silicate		–		–		–		–		–		–		–		–	13	1379
Glass, sodium silicate no. 23		–	5	1240		–		–		–		–		–		–		–
Glass, soft	2	511		–		–		–		–		–		–		–		–
Glass, solex 2808 plate	2	923		–		–		–		–		–		–		–		–
Glass, solex 2808X plate	2	925	5	1240		–		–		–		–		–		–		–
Glass, solex S	2	925		–		–		–		–		–		–		–		–
Glass, solex S plate	2	923	5	1240		–		–		–		–		–		–		–
Glass, strontium borate		–		–		–		–		–		–		–		–	13	1380
Glass, synthetic tektite		–		–		–		–		–		–	10	579		–		–
Glass, television tube		–		–	8	1644	8	1646		–	8	1648		–		–		–
Glass, thuringian	2	923 924		–		–		–		–		–		–		–		–
Glass, vycor, high silica		–	5	1234		–		–		–		–		–		–		–
Glass, vycor 790		–		–		–		–		–	8	1633		–		–		–
Glass, vycor 791		–		–		–		–		–	8	1633		–		–		–
Glass, vycor 7900		–	5	1324		–		–		–		–		–		–		–
Glass, vycor-brand	2	926		–		–		–		–		–		–		–		–
Glass, vycor-brand No. 790		–		–		–		–		–		–		–		–	13	1356
Glass, white plate	2	923 925		–		–		–		–		–		–		–		–
Glass, window	2	923 924		–		–		–		–		–		–		–		–
Glass, x-ray protection	2	924		–		–		–		–		–		–		–		–
Glass-ceramic, BDQ 115		–		–		–		–		–		–	10	583		–		–
Glass fiber, blanket superfine	2	1116		–		–		–		–		–		–		–		–
Glass fiber, insulation blanket	2	1117		–		–		–		–		–		–		–		–
Glasses, miscellaneous		–		–	8	1643	8	1645		–	8	1647 1649		–		–		–
Glaubertie		–		–		–		–		–	8	1694		–		–		–
Globar, silicon monocarbide, SiC		–		–	8	798		–	8	808 810		–		–		–		–
Glucinum	1	18	4	16		–		–		–		–		–		–		–
Glucinum sulfate		–	5	1179		–		–		–		–		–		–		–
Glycerin		–	6	230		–		–		–		–		–		–		–
Glycerol	3	209	6	230		–		–		–		–	10	589		–		–
Glycerol tribromohydrin		–	6s	91		–		–		–		–		–		–		–
Glycerol trichlorohydrin		–	6s	91		–		–		–		–		–		–		–
Glycol		–	6	192		–		–		–		–		–		–		–
Glycol dibromide		–	6s	26		–		–		–		–		–		–		–
Glycol dichloride		–	6s	27		–		–		–		–		–		–		–
Glycyl alcohol		–	6	230		–		–		–		–		–		–		–
Goethite, ore		–		–		–	8	1676		–	8	1661		–		–		–
Gold, Au	1	132	4	83	7	244	7	258	7	269		–	10	73		–	12	125

Substance Name	Thermal Conductivity		Specif. Heat		Thermal Radiative Properties								Thermal Diffusivity		Viscosity		Thermal Expansion	
					Emissivity		Reflectivity		Absorptivity		Transmissiv.							
	V.	Page	V.	Page	V.	Page	V.	Page	V.	Page	V.	Page	V.	Page	V.	Page	V.	Page
Gold, Au (continued)	1	132	4	83	7	244 248 250 254	7	258 264 267	7	269 271 273 275 277	–		10	73	–		12	125
Gold, 5 percent impurities	–		–		–		–		–		–		10	289	–		–	
Gold, mint	–		–		–		–		–		–		10	289	–		–	
Gold alloys:																		
Au + Ag	1	620	–		–		7	932	–		–		–		–		12	821
Au + Cd	1	600	–		–		–		–		–		–		–		12	688
Au + Co	1	606	–		–		–		–		–		–		–		–	
Au + Cr	1	603	–		–		–		–		–		–		–		–	
Au + Cu	1	609	–		–		–		–		–		–		–		12	763
Au + In	–		–		–		–		–		–		–		–		12	809
Au + Ni	–		4	353	–		–		–		–		–		–		–	
Au + Pd	1	614	–		–		–		–		–		–		–		12	812
Au + Pd, palau	–		–		–		–		–		–		–		–		12	814
Au + Pd, pallagold	–		–		–		–		–		–		–		–		12	814
Au + Pt	1	617	–		–		–		–		–		–		–		12	817
Au + Zn	1	623	–		–		–		–		–		–		–		–	
Au + Ag + ΣX_i	–		–		–		–		–		–		–		–		12	1121
Gold-aluminum intermetallic compound, $AuAl_2$	–		–		–		8	1295	–		–		–		–		–	
Gold-copper intermetallic compounds:																		
CuAu	1	1282	–		–		–		–		–		–		–		–	
Cu_3Au	1	1282	–		–		–		–		–		–		–		–	
Gold-gallium intermetallic compound, $AuGa_2$	–		–		–		8	1295	–		–		–		–		–	
Gold-indium intermetallic compound, $AuIn_2$	–		–		–		8	1295	–		–		–		–		–	
Gold-manganese intermetallic compounds:																		
AuMn	–		–		–		–		–		–		–		12	550		
Au_2Mn	–		–		–		–		–		–		–		12	547		
Au_4Mn	–		–		–		–		–		–		–		12	546		
Gold-vanadium intermetallic compound, Au_4V	–		–		–		–		–		–		–		12	552		
Gold-zinc intermetallic compound, AuZn	–		–		–		–		–		–		–		12	554		
Granite	2	817	–		–		–		–		–		–		–		–	
Granite, NTS	2	818	–		–		–		–		–		–		–		–	
Granite + blast furnace slag	–		–		–		–		–		–		10	438	–		–	
Grapefruit	–		–		–		–		–		–		10	636	–		–	
Graphite, 49 B-2	–		–		–		–		–		–		–		–		13	129
Graphite, 50 B-1	–		–		–		–		–		–		–		–		13	130
Graphite, 50 D-1	–		–		–		–		–		–		–		–		13	130
Graphite, 50 I-1	–		–		–		–		–		–		–		–		13	130
Graphite, 580	–		–		8	33 46	–		–		–		–		–		–	

Substance Name	Thermal Conductivity		Specif. Heat		Thermal Radiative Properties								Thermal Diffusivity		Viscosity		Thermal Expansion	
					Emissivity		Reflectivity		Absorptivity		Transmissiv.							
	V.	Page	V.	Page	V.	Page	V.	Page	V.	Page	V.	Page	V.	Page	V.	Page	V.	Page
Graphite, British, reactor grade carbon	2	69 70		–		–		–		–		–		–		–		–
Graphite, brom-graphite	2	768		–		–		–		–		–		–		–		–
Graphite, brookhaven	2	26		–		–		–		–		–		–		–		–
Graphite, C		–		–		–		–		–		–		–		–	13	52
Graphite, CA		–		–		–		–		–		–		–		–	13	124
Graphite, Canadian natural	2	54	5	9 11		–		–		–		–		–		–		–
Graphite, carbon		–		–		–		–		–		–		–		–	13	24
Graphite, carbon resistor	2	73		–		–		–		–		–		–		–		–
Graphite, CB		–		–		–		–		–		–		–		–	13	124
Graphite, CBN		–		–		–		–		–		–		–		–	13	124
Graphite, CDG	2	65		–		–		–		–		–		–		–		–
Graphite, CEP		–		–		–		–	8	75		–	10	32 33 37		–	13	124
Graphite, CEQ	2	63 65		–		–		–		–		–		–		–		–
Graphite, ceylon		–		–		–		–		–		–		–		–	13	129
Graphite, ceylon natural		–	5	9		–		–		–		–		–		–		–
Graphite, CFW	2	67		–		–		–		–		–		–		–	13	56
Graphite, CFZ	2	67 71 72		–		–		–		–		–		–		–	13	59
Graphite, CS	2	54 55 56 64	5	9		–		–		–		–	10	29 36 37		–		–
Graphite, CS112	2	63		–		–		–		–		–		–		–		–
Graphite, CS312	2	63		–		–		–		–		–		–		–	13	124
Graphite, CSF	2	55		–		–		–		–		–		–		–		–
Graphite, CSF-MTR	2	63		–		–		–		–		–		–		–		–
Graphite, deposited carbon	2	32		–		–		–		–		–		–		–		–
Graphite, EH		–		–		–		–		–		–		–		–	13	125
Graphite, expanded pyrolytic		–		–		–		–		–		–	10	37		–		–
Graphite, experimental grade		–		–		–		–		–		–		–		–	13	64
Graphite, EY 9	2	69 70 71		–		–		–		–		–		–		–	13	125
Graphite, EY 9A	2	70		–		–		–		–		–		–		–		–
Graphite, fuel-filled	2	545 548 558		–		–		–		–		–		–		–		–
Graphite, G-5	2	60 61		–		–		–		–		–		–		–		–
Graphite, G-9	2	60 61		–		–		–		–		–		–		–	13	72 73 74
Graphite, GBE	2	54 55		–	8	32 45	8	66	8	76		–		–		–	13	68
Graphite, GBH	2	55	5	11	8	33 45	8	66	8	76		–		–		–	13	71

Substance Name	Thermal Conductivity		Specif Heat		Thermal Radiative Properties								Thermal Diffusivity		Viscosity		Thermal Expansion	
					Emissivity		Reflectivity		Absorptivity		Transmissiv.							
	V.	Page	V.	Page	V.	Page	V.	Page	V.	Page	V.	Page	V.	Page	V.	Page	V.	Page
Graphite, GLI–S4, Great Lakes impervious		–		–		–		–		–		–		–		–	13	125
Graphite, H1LM		–		–	8	33 46		–		–		–		–		–		–
Graphite, H3LM		–		–	8	33 46		–		–		–		–		–	13	125
Graphite, H4LM	2	61		–		–		–		–		–		–		–		–
Graphite, H205		–		–		–		–		–		–		–		–	13	125
Graphite, H249		–		–		–		–		–		–		–		–	13	125
Graphite, Japan domestic	2	56		–		–		–		–		–		–		–		–
Graphite, JTA	2	70 72		–		–		–		–		–		–		–	13	126
Graphite, karbate	2	59		–		–		–		–		–		–		–		–
Graphite, Korite	2	55		–		–		–		–		–		–		–		–
Graphite, L–117	2	63		–		–		–		–		–		–		–		–
Graphite, MHLM		–		–		–		–		–		–		–		–	13	126
Graphite, MH4LM	2	70		–		–		–		–		–		–		–		–
Graphite, moderator	2	70		–		–		–		–		–		–		–		–
Graphite, natural Ceylon	2	55		–		–		–		–		–		–		–	13	126
Graphite, natural Madagascan		–	5	9		–		–		–		–		–		–		–
Graphite, ohmite	2	73		–		–		–		–		–		–		–		–
Graphite, P1	2	35		–		–		–		–		–		–		–		–
Graphite, P–03		–		–		–		–		–		–		–		–	13	126
Graphite, pencil lead	2	65		–		–		–		–		–		–		–		–
Graphite, PGX		–		–		–		–		–		–		–		–	13	126
Graphite, pile H–CSII		–	5	9		–		–		–		–		–		–		–
Graphite, POCO		–		–		–		–		–		–		–		–	13	75
Graphite, porous–40	2	63		–		–		–		–		–		–		–		–
Graphite, porous–60	2	63		–		–		–		–		–		–		–		–
Graphite, pyro		–	5	9		–		–		–		–		–		–		–
Graphite, pyrolytic	2	30		–	8	30 33 53	8	66 70		–		–	10	32 37		–	13	79
Graphite, pyrolytic, expanded		–		–		–		–		–		–	10	37		–		–
Graphite, pyrolytic, filament	2	32		–		–		–		–		–		–		–		–
Graphite, pyrolytic, supertemp		–		–		–		–		–		–	10	33		–		–
Graphite, R0008	2	60		–		–		–		–		–		–		–		–
Graphite, R0025	2	71		–		–		–		–		–		–		–		–
Graphite, reactor grade		–		–		–		–		–		–	10	37		–		–
Graphite, reactor grade carbon stock	2	73		–		–		–		–		–		–		–		–
Graphite, RT0003	2	54		–		–		–		–		–		–		–		–
Graphite, RT0029		–		–		–		–		–		–		–		–	13	89
Graphite, RVA	2	66 67		–		–		–		–		–		–		–	13	93 94 96
Graphite, RVC		–		–		–		–		–		–		–		–	13	97

Substance Name	Thermal Conductivity		Specif. Heat		Thermal Radiative Properties								Thermal Diffusivity		Viscosity		Thermal Expansion	
					Emissivity		Reflectivity		Absorptivity		Transmissiv.							
	V.	Page	V.	Page	V.	Page	V.	Page	V.	Page	V.	Page	V.	Page	V.	Page	V.	Page
Graphite, RVD	2	67	–		–		–		–		–		10	38	–		13	101
Graphite, SA25	2	42	–		–		–		–		–		–		–		–	
Graphite, siliconized		–	–		9	1325 1326 1328	–		–		–		–		–		–	
Graphite, SPK		–	–		8	58	8	70 74	–		–		–		–		–	
Graphite, supertemp pyrolytic	2	72	–		–		–		–		–		10	33	–		–	
Graphite, SX-5		–	–		–		–		–		–		10	37	–		13	127
Graphite, thermax W		–	–		–		–		–		–		10	30 36	–		–	
Graphite, TS-148	2	59	–		–		–		–		–		–		–		–	
Graphite, TS-160	2	59	–		–		–		–		–		–		–		–	
Graphite, TS-574		–	–		–		–		–		–		–		–		13	127
Graphite, TS-699		–	–		–		–		–		–		10	36	–		–	
Graphite, TS-835		–	–		–		–		–		–		–		–		13	127
Graphite, TSP Nuclear grade	2	60	–		–		–		–		–		–		–		–	
Graphite, TSX		–	–		–		–		–		–		–		–		13	127
Graphite, U. B. carbon A		–	–		–		–		–		–		10	30 35	–		–	
Graphite, U. B. carbon R		–	–		–		–		–		–		10	30 35	–		–	
Graphite, U. B. carbon Z		–	–		–		–		–		–		10	30 36	–		–	
Graphite, U. B. A		–	–		–		–		–		–		10	31	–		–	
Graphite, U. B. G		–	–		–		–		–		–		10	31	–		–	
Graphite, U. B. R		–	–		–		–		–		–		10	31	–		–	
Graphite, U. B. Z		–	–		–		–		–		–		10	31	–		–	
Graphite, UT6		–	–		–		8	64	–		–		–		–		–	
Graphite, W		–	–		–		–		–		–		–		–		13	128
Graphite, W, Specialties Co.		–	–		–		–		–		–		10	31	–		–	
Graphite, ΩV5G Great Lakes		–	–		–		–		–		–		–		–		13	125
Graphite, ZT	2	60 61 71	–		–		–		–		–		–		–		–	
Graphite, ZTA	2	65 66 70	–		–		–		–		–		–		–		13	107
Graphite, ZTB		–	–		–		–		–		–		–		–		13	115
Graphite, ZTC	2	66	–		–		–		–		–		–		–		13	119
Graphite, ZTD	2	66	–		–		–		–		–		–		–		13	128
Graphite, ZTE	2	66	–		–		–		–		–		–		–		13	128
Graphite, ZTF	2	66	–		–		–		–		–		–		–		13	129
Graphite + bromine, mixture	2	767	–		–		–		–		–		–		–		–	
Graphite nitrate, $C_{24}NO_3$		–	–		–		–		–		–		–		–		13	671
Graphite + silicon carbide, mixture	2	789	–		–		–		–		–		–		–		–	
Graphite + tantalum carbide, mixture		–	–		–		–		–		–		–		–		13	951
Graphite + thorium dioxide, mixture	2	544 557	–		–		–		–		–		–		–		–	

| Substance Name | Thermal Conductivity | | Specif. Heat | | Thermal Radiative Properties | | | | | | | | Thermal Diffusivity | | Viscosity | | Thermal Expansion | |
| | | | | | Emissivity | | Reflectivity | | Absorptivity | | Transmissiv. | | | | | | | |
	V.	Page	V.	Page	V.	Page	V.	Page	V.	Page	V.	Page	V.	Page	V.	Page	V.	Page
Graphite + uranium dicarbide, mixture	2	770		–		–		–		–		–		–		–		–
Graphite + uranium dioxide, mixture	2	547		–		–		–		–		–		–		–		–
Graphite + zirconium carbide, mixture		–		–		–		–		–		–		–		–	13	964
Gypsum		–		–		–	8	1698		–		–		–		–		–
Hafnates		–		–		–	8	596		–		–		–		–		–
Hafnia	2	150	5	101		–		–		–		–		–		–		–
Hafnium, Hf	1	138	4	87	7	280 282 284		–		–		–	10	77		–	12	134
Hafnium alloys:																		
Hf + Zr	1	624	4	356		–		–		–		–	10	242		–	12	822
Hf + Ta + ΣXi		–		–		–		–		–		–	10	290		–		–
Hafnium beryllide, Hf₂Be₂₁		–	5	313		–		–		–		–		–		–		–
Hafnium boride, HfB₂		–	5	341	8	730 732		–		–		–	10	465		–	13	758
Hafnium carbide, HfC	2	575	5	420	8	850 852		–		–		–	10	467		–	13	848
Hafnium carbide + carbon, mixture		–		–		–		–		–		–		–		–	13	946
Hafnium diboride + carbon, mixture		–		–		–		–		–		–	10	521		–		–
Hafnium diboride + silicon carbide, mixture		–		–		–		–		–		–	10	523		–		–
Hafnium diboride + silicon carbide + carbon, mixture		–		–		–		–		–		–	10	525		–		–
Hafnium dioxide + iron, cermet		–		–		–		–		–		–		–		–	13	1317
Hafnium fluoride, HfF₄		–	5	937		–		–		–		–		–		–		–
Hafnium nitride, HfN	2	659	5	1081	8	1056 1058 1060		–		–		–		–		–	13	1162
Hafnium oxide, HfO₂	2	150	5	101	8	273	8	275		–		–		–		–	13	251
Hafnium silicate, HfSiO₄		–		–		–		–		–		–		–		–	13	727
Hafnium tantalum oxide, 6HfO₂·Ta₂O₅		–		–		–		–		–		–		–		–	13	534
Hair felt	2	1099		–		–		–		–		–		–		–		–
Ham		–		–		–		–		–		–	10	639		–		–
Haynes LT-1, cermet		–		–	8	1356		–		–		–		–		–		–
Haynes LT-1B, cermet		–		–	8	1356		–		–		–		–		–		–
Haynes LT-2, cermet		–		–	8	1375		–		–		–		–		–		–
Haynes stellite alloy 21, Vitallium type alloy	1	948		–		–		–		–		–		–		–		–
Heavy ethane, C₂D₆		–	6s	35		–		–		–		–		–		–		–
Heavy hydrogen, D₂	3	21	6	15		–		–		–		–		–		–		–
Heavy hydrogen, monatomic		–	6s	26		–		–		–		–		–		–		–
Heavy water, D₂O		–	6s	95		–		–		–		–		–		–		–
Helium, He	3	29	6	23		–		–		–		–		–	11	18		–
Helium-hydrogen, mixture	3	333		–		–		–		–		–		–	11	302		–
Helium-krypton, mixture	3	276		–		–		–		–		–		–	11	260		–
Helium-krypton-xenon, mixture	3	480		–		–		–		–		–		–		–		–
Helium-methane, mixture	3	338		–		–		–		–		–		–		–		–

Substance Name	Thermal Conductivity		Specif. Heat		Thermal Radiative Properties								Thermal Diffusivity		Viscosity		Thermal Expansion	
					Emissivity		Reflectivity		Absorptivity		Transmissiv.							
	V.	Page	V.	Page	V.	Page	V.	Page	V.	Page	V.	Page	V.	Page	V.	Page	V.	Page
Helium-neon, mixture	3	271	–		–		–		–		–			–	11	269	–	
Helium-neon-deuterium, mixture	3	489	–		–		–		–		–			–		–		–
Helium-neon-xenon, mixture	3	482	–		–		–		–		–			–		–		–
Helium-nitrogen, mixture	3	340	–		–		–		–		–			–	11	308	–	
Helium-nitrogen-methane, mixture	3	487	–		–		–		–		–			–		–		–
Helium-oxygen, mixture	3	343	–		–		–		–		–			–	11	322	–	
Helium-oxygen-methane, mixture	3	484	–		–		–		–		–			–		–		–
Helium-propane, mixture	3	345	–		–		–		–		–			–		–		–
Helium-propylene, mixture	3	347	–		–		–		–		–			–		–		–
Helium-xenon, mixture	3	280	–		–		–		–		–			–	11	277	–	
Hematite		–		–	–		8	1677	–		–			–		–		–
Hematite, oolitic		–		–	–		8	1678	–		–			–		–		–
Hemiterpene		–	6s	56	–		–		–		–			–		–		–
n-heptane	3	211	6	232	–		–		–		–			–	11	182	–	
n-heptane-nitrogen, mixture		–		–	–		–		–		–			–	11	436	–	
Heulandite		–		–	–		–		–		8	1694		–		–		–
Hexadecafluoro-n-heptane-2,2,4-trimethylpentane		–		–	–		–		–		–			–	11	438	–	
Hexadecane		–	6s	43	–		–		–		–			–		–		–
Hexadeuteriobenzene		–	6s	2	–		–		–		–			–		–		–
Hexadeuterioethane		–	6s	35	–		–		–		–			–		–		–
Hexafluoroethane, R116		–	6s	44	–		–		–		–			–		–		–
Hexahydrobenzene		–	6s	25	–		–		–		–			–		–		–
Hexahydrotoluene		–	6s	62	–		–		–		–			–		–		–
Hexamethylbenzene		–	6s	44	–		–		–		–			–		–		–
Hexamethylene		–	6s	25	–		–		–		–			–		–		–
Hexane, n-C_6H_{14}	3	214		–	–		–		–		–			–		–		–
Hexane-methanol, mixture	3	460	6	238	–		–		–		–			–	11	184	–	
1-Hexanol		–	6s	44	–		–		–		–			–		–		–
Hexone		–	6s	66	–		–		–		–			–		–		–
Hexyl alcohol		–	6s	44	–		–		–		–			–		–		–
Holmia		–	5	104	–		–		–		–			–		–		–
Holmium, Ho	1	142	4	90	–		–		–		–		10	78		–	12	138
Holmium borides:																		
HoB$_6$		–		–	–		–		–		8	727		–		–		–
HoB$_{12}$		–		–	–		–		–		–			–		–	13	763
Holmium carbide, HoC_2		–		–	–		–		–		–			–		–	13	935
Holmium chloride hexahydrate, $HoCl_3 \cdot 6H_2O$		–	5	829	–		–		–		–			–		–		–
Holmium nitride, HoN		–		–	–		–		–		8	1088		–		–		–
Holmium oxide, Ho_2O_3		–	5	104	–		–		–		8	277		–		–	13	257
Holmium-zinc intermetallic compound, $HoZn_2$		–		–	–		–		–		–			–		–	12	557
Honeycomb structures, metallic-nonmetallic	2	1015		–	–		–		–		–			–		–		–

Substance Name	Thermal Conductivity		Specif. Heat		Emissivity		Reflectivity		Absorptivity		Transmissiv.		Thermal Diffusivity		Viscosity		Thermal Expansion	
	V.	Page	V.	Page	V.	Page	V.	Page	V.	Page	V.	Page	V.	Page	V.	Page	V.	Page
Honeycomb structures, nonmetallic	2	1010		–		–		–		–		–		–		–		–
Hydrargillite		–		–		–		–		–	8	1664		–		–		–
Hydrargyrum	1	212	4	131		–		–		–		–		–		–		–
Hydrazine		–	6s	44		–		–		–		–		–		–		–
Hydrazine, anhydrous		–	6s	44		–		–		–		–		–		–		–
Hydrobromic acid		–	6s	45		–		–		–		–		–		–		–
Hydrobromic ether		–	6s	4		–		–		–		–		–		–		–
Hydrobromide		–	6s	45		–		–		–		–		–		–		–
Hydrochinone		–	6s	53		–		–		–		–		–		–		–
Hydrochloric acid, HCl	3	101		–		–		–		–		–		–		–		–
Hydrocyanic acid		–	6s	46		–		–		–		–		–		–		–
Hydrofluoric acid		–	6s	46		–		–		–		–		–		–		–
Hydrofluoric acid, monodeuterated		–	6s	47		–		–		–		–		–		–		–
Hydrogen, H_2	3	41		–		–		–		–		–		–		–		–
Hydrogen, monodeuterated		–	6s	48		–		–		–		–		–		–		–
Hydrogen, sulfuretted		–	6	78		–		–		–		–		–		–		–
Hydrogen arsenide		–	6s	2		–		–		–		–		–		–		–
Hydrogen bromide		–	6s	45		–		–		–		–		–		–		–
Hydrogen chloride, HCl	3	101	6	72		–		–		–		–		–	11	76		–
Hydrogen cyanide		–	6s	46		–		–		–		–		–		–		–
Hydrogen fluoride, HF		–	6s	46		–		–		–		–		–		–		–
Hydrogen fluoride, monodeuterated		–	6s	47		–		–		–		–		–		–		–
Hydrogen–hydrogen chloride, mixture		–		–		–		–		–		–		–	11	521		–
Hydrogen–hydrogen deuteride, mixture		–		–		–		–		–		–		–	11	440		–
Hydrogen iodide, HI	3	103	6	76		–		–		–		–		–	11	78		–
Hydrogen–krypton, mixture	3	351		–		–		–		–		–		–		–		–
Hydrogen–methane, mixture		–				–		–		–		–		–	11	442		–
Hydrogen–methane–nitrogen, mixture		–		–		–		–		–		–		–	11	587		–
Hydrogen monatomic		–	6s	48		–		–		–		–		–		–		–
Hydrogen–neon, mixture	3	362		–		–		–		–		–		–	11	337		–
Hydrogen–neon–nitrogen, mixture	3	494		–		–		–		–		–		–		–		–
Hydrogen–neon–oxygen, mixture	3	492		–		–		–		–		–		–		–		–
Hydrogen–nitric oxide, mixture		–		–		–		–		–		–		–	11	445		–
Hydrogen–nitrogen, mixture	3	419		–		–		–		–		–		–	11	447		–
Hydrogen–nitrogen–oxygen, mixture	3	498		–		–		–		–		–		–		–		–
Hydrogen–nitrous oxide, mixture	3	427		–		–		–		–		–		–	11	458		–
Hydrogen dioxide, H_2O_2		–	6s	49		–		–		–		–		–		–		–
Hydrogen peroxide, H_2O_2		–	6s	49		–		–		–		–		–		–		–
Hydrogen–oxygen, mixture	3	429		–		–		–		–		–		–	11	460		–
Hydrogen phosphide		–	6s	74		–		–		–		–		–		–		–
Hydrogen–propane, mixture		–		–		–		–		–		–		–	11	463		–
Hydrogen selenide		–	6s	49		–		–		–		–		–		–		–

Substance Name	Thermal Conductivity		Specif. Heat		Thermal Radiative Properties								Thermal Diffusivity		Viscosity		Thermal Expansion	
					Emissivity		Reflectivity		Absorptivity		Transmissiv.							
	V.	Page	V.	Page	V.	Page	V.	Page	V.	Page	V.	Page	V.	Page	V.	Page	V.	Page
Hydrogen selenide, dideuterated		–	6s	49		–		–		–		–		–		–		–
Hydrogen sulfide, H₂S	3	104	6	78		–		–		–		–		–	11	80		–
Hydrogen sulfide, dideuterated		–	6s	50		–		–		–		–		–		–		–
Hydrogen sulfide, ditritiated		–	6s	51		–		–		–		–		–		–		–
Hydrogen sulfide, monodeuterated		–	6s	51		–		–		–		–		–		–		–
Hydrogen sulfide, monodeuterated, monotritiated		–	6s	52		–		–		–		–		–		–		–
Hydrogen sulfide, monotritiated		–	6s	52		–		–		–		–		–		–		–
Hydrogen-sulfur dioxide, mixture		–		–		–		–		–		–		–	11	523		–
Hydrogen tritium sulfide		–	6s	52		–		–		–		–		–		–		–
Hydrogen-xenon, mixture	3	374		–		–		–		–		–		–		–		–
Hydriodic acid, HI	3	103		–		–		–		–		–		–		–		–
Hydroquinol		–	6s	53		–		–		–		–		–		–		–
Hydroquinone		–	6s	53		–		–		–		–		–		–		–
Hydroxyacetanilide		–	6s	53		–		–		–		–		–		–		–
Hydroxyl		–	6s	53		–		–		–		–		–		–		–
1-Hydroxynaphthalene		–	6s	69		–		–		–		–		–		–		–
2-Hydroxynaphthalene		–	6s	69		–		–		–		–		–		–		–
3-Hydroxyphenol		–	6s	83		–		–		–		–		–		–		–
α-Hydroxytoluene		–	6s	2		–		–		–		–		–		–		–
Hypo	2	693		–		–		–		–		–		–		–		–
Ice		–		–		–		–		–		–	10	390		–	13	261
Illinium	1	285		–		–		–		–		–		–		–		–
Indium In	1	146	4	95		–		–	7	286		–	10	79		–	12	143
Indium alloys:																		
In + Pb	1	627		–		–		–		–		–		–		–	12	827
In + Sn	1	634	4	359		–		–		–		–		–		–	12	835
In + Tl	1	630		–		–		–		–		–		–		–	12	832
Indium antimonide, InSb	1	1287	5	303		–	8	1298		–	8	1305	10	370		–		–
Indium antimonide + gallium antimonide, mixture		–		–		–	8	1297		–		–		–				–
Indium antimonide + indium telluride, mixture	1	1403		–		–		–		–		–		–				–
Indium-antimony intermetallic compound, InSb	1	1287		–		–		–		–		–		–				–
Indium arsenide, InAs	1	1292	5	310	8	685	8	687		–	8	689		–		–	13	752
Indium arsenide + indium phosphide, mixture	1	1426		–		–		–		–		–		–		–		–
Indium gallium phosphide, InGaP		–		–		–		–		–		–		–		–	13	1183
Indium oxides:																		
InO	2	153		–		–		–		–		–		–		–		–
In₂O₃		–		–		–	8	546		–		–		–		–	13	267
Indium-palladium intermetallic compound, In₃Pd		–		–		–		–		–		–		–		–	12	561 563 564

| Substance Name | Thermal Conductivity | | Specif. Heat | | Thermal Radiative Properties | | | | | | | | Thermal Diffusivity | | Viscosity | | Thermal Expansion | |
| | | | | | Emissivity | | Reflectivity | | Absorptivity | | Transmissiv. | | | | | | | |
	V.	Page	V.	Page	V.	Page	V.	Page	V.	Page	V.	Page	V.	Page	V.	Page	V.	Page
Indium phosphide, InP		–	5	523		–	8	1100		–	8	1103		–		–	13	1183
Indium-praseodymium intermetallic compound, In₃Pr		–		–		–		–		–		–		–		–	12	562 563 565
Indium selenide, In₂Se₃	1	1295		–		–		–		–		–		–		–		–
Indium sulfide, nonstoichiometric		–	5	668		–		–		–		–		–		–		–
Indium telluride, In₂Te₃	1	1298		–		–		–		–		–		–		–	13	1270
Indium-ytterbium intermetallic compound, In₃Yb		–		–		–		–		–		–		–		–	12	566 568 569
Indium-yttrium intermetallic compound, In₃Y		–		–		–		–		–		–		–		–	12	567 568 570
Intermetallic compounds:																		
AgCd		–		–		–	8	1326		–		–		–		–		–
AgCu	1	1338		–		–		–		–		–		–		–		–
AgSbTe₂	1	1335		–		–		–		–		–		–		–		–
AgZn		–		–		–	8	1328		–		–		–		–	12	619
Ag₂Al		–		–		–	8	1352		–		–		–		–		–
Ag₂Tb		–		–		–		–		–		–		–		–	12	618
AlAg₂		–		–		–		–		–		–		–		–	12	444
Al₂Au		–		–		–		–		–		–		–		–	12	423
AlCu		–		–		–		–		–		–		–		–	12	417 419
AlCu₂		–		–		–		–		–		–		–		–	12	417 420
AlCu₃		–		–		–		–		–		–		–		–	12	417 422
Al₂Cu		–		–		–		–		–		–		–		–	12	417 418
Al₄Cu₉		–		–		–		–		–		–		–		–	12	417 421
AlFe		–		–		–		–		–		–		–		–	12	433 435 436
AlFe₃		–		–		–		–		–		–		–		–	12	434 435 437
Al₂Fe		–		–		–		–		–		–		–		–	12	428 429 432
Al₃Fe		–		–		–		–		–		–		–		–	12	427 429 431
Al₅Fe₂		–		–		–		–		–		–		–		–	12	426 429 430
AlNi		–		–		–		–		–		–		–		–	12	438 440
AlNi₃		–		–		–		–		–		–		–		–	12	441 443
AlSb		–		–		–		–		–		–		–		–	12	414

Substance Name	Thermal Conductivity		Specif. Heat		Thermal Radiative Properties											Thermal Diffusivity		Viscosity		Thermal Expansion	
					Emissivity		Reflectivity		Absorptivity		Transmissiv.											
	V.	Page	V.	Page	V.	Page	V.	Page	V.	Page	V.	Page			V.	Page	V.	Page	V.	Page		
Intermetallic compounds: (continued)																						
Al_3U		–		–		–		–		–		–				–		–	12	447		
$AuAl_2$		–		–		–	8	1295		–		–				–		–		–		
$AuGa_2$		–		–		–	8	1295		–		–				–		–		–		
$AuIn_2$		–		–		–	8	1295		–		–				–		–		–		
$AuMn$		–		–		–		–		–		–				–		–	12	550		
Au_2Mn		–		–		–		–		–		–				–		–	12	547		
Au_4Mn		–		–		–		–		–		–				–		–	12	546		
Au_4V		–		–		–		–		–		–				–		–	12	552		
$AuZn$		–		–		–	8	1292		–		–				–		–	12	554		
Ba_2Pb	1	1245		–		–		–		–		–				–		–		–		
Ba_2Sn	1	1246		–		–		–		–		–				–		–		–		
Be_2Cr		–		–	8	1275		–		–		–				–		–		–		
$Be_{12}Nb$	1	1248		–	8	1273	8	1280		–		–				–		–	12	464		
$Be_{17}Nb_2$	1	1248		–	8	1273 1277	8	1280		–		–				–		–		–		
$\beta\text{-}Be_{17}Hf_2$		–		–		–		–		–		–				–		–	12	461		
Be_2Re		–		–	8	1275		–		–		–				–		–		–		
$Be_{13}Sc$		–		–	8	1275		–		–		–				–		–		–		
$Be_{12}Ta$	1	1251		–	8	1273 1277	8	1280		–		–				–		–	12	467 469 470		
$Be_{17}Ta_2$	1	1251		–	8	1273 1277	8	1280		–		–				–		–	12	468 469 471		
Be_2Ti		–		–	8	1275		–		–		–				–		–		–		
$Be_{13}U$	1	1254		–		–		–		–		–				–		–	12	472		
$Be_{13}Zr$	1	1256		–		–	8	1280		–		–				–		–	12	475		
$Be_{17}Zr_2$		–		–		–	8	1280		–		–				–		–		–		
$BiPt$		–		–		–		–		–		–				–		–	12	479 480 482		
Bi_2Pt		–		–		–		–		–		–				–		–	12	478 480 481		
$CaMg_2$		–		–		–		–		–		–				–		–	12	493		
Ca_2Pb	1	1271		–		–		–		–		–				–		–		–		
Ca_2Sn	1	1273		–		–		–		–		–				–		–		–		
$CdAu$		–		–		–		–		–		–				–		–	12	483		
$CdSb$	1	1264		–		–		–		–		–				–		–		–		
$CdLi$		–		–		–		–		–		–				–		–	12	487 489 490		
Cd_3Mg		–		–		–		–		–		–				–		–	12	489 491		
$CeIn_3$		–		–		–		–		–		–				–		–	12	496 498 499		
$CePd_3$		–		–		–		–		–		–				–		–	12	497 498 500		

Substance Name	Thermal Conduc- tivity		Specif. Heat		Thermal Radiative Properties										Thermal Diffu- sivity		Visco- sity		Thermal Expan- sion	
					Emis- sivity		Reflec- tivity		Absorp- tivity		Trans- missiv.									
	V.	Page	V.	Page	V.	Page	V.	Page	V.	Page	V.	Page	V.	Page	V.	Page	V.	Page
Intermetallic compounds: (continued)																		
CeRu₂	-		-		-		-		-		-		-		-	12	501
CeSn₃	-		-		-		-		-		-		-		-	12	504
CoAl	-		-		-		8	1352	-		-		-		-		-
Co₂Dy	-		-		-		-		-		-		-		-	12	508 510 511
Co₂Gd	-		-		-		-		-		-		-		-	12	509 510 512
Co₅Y	-		-		-		-		-		-		-		-	12	514 517
Co₁₇Y₂	-		-		-		-		-		-		-		-	12	513 515 516
CrFe	-		-		-		-		-		-		-		-	12	507
CuAu	1	1282	-		-		-		-		-		-		-	12	519 520 522
Cu₃Au	1	1282	-		-		-		-		-		-		-		-
Cu₂Mg	-		-		-		-		-		-		-		-	12	523
CuSbSe₂	1	1275	-		-		-		-		-		-		-		-
Cu₄Sn	-		-		-		8	1352	-		-		-		-		-
CuZn	-		-		-		8	1285	-		-		-		-	12	525
Cu₅Zn₆	-		-		-		-		-		-		-		-	12	524
Fe₁₇Lu₂	-		-		-		-		-		-		-		-	12	571
FeNi₃	-		-		-		-		-		-		-		-	12	574
FeRh	-		-		-		-		-		-		-		-	12	575
Fe₁₇Y₂	-		-		-		-		-		-		-		-	12	576
GaAg₃	-		-		-		-		-		-		-		-	12	536
GaNi	-		-		-		-		-		-		-		-	12	533
Gd₂Fe₁₇	-		-		-		-		-		-		-		-	12	527 529 531
GdIn₃	-		-		-		-		-		-		-		-	12	526 529 530
GdPd₃	-		-		-		-		-		-		-		-	12	528 529 532
GeLa	-		-		-		-		-		-		-		-	12	537
GeMg₂	-		-		-		-		-		-		-		-	12	538
GePr	-		-		-		-		-		-		-		-	12	540 542 544
Ge₂Pr	-		-		-		-		-		-		-		-	12	539 542 543
Ge₃Pr₅	-		-		-		-		-		-		-		-	12	541 542 545
HoZn₂	-		-		-		-		-		-		-		-	12	557

Substance Name	Thermal Conductivity		Specif. Heat		Thermal Radiative Properties										Thermal Diffusivity		Viscosity		Thermal Expansion	
					Emissivity		Reflectivity		Absorptivity		Transmissiv.									
	V.	Page	V.	Page	V.	Page	V.	Page	V.	Page	V.	Page	V.	Page	V.	Page	V.	Page
Intermetallic compounds: (continued)																		
In$_3$Pd		–		–		–		–		–		–		–		–	12	561 563 564
In$_3$Pr		–		–		–		–		–		–		–		–	12	562 563 565
InSb	1	1287		–		–		–		–		–		–		–		–
In$_3$Y		–		–		–		–		–		–		–		–	12	567 568 570
In$_3$Yb		–		–		–		–		–		–		–		–	12	566 568 569
LaRu$_2$		–		–		–		–		–		–		–		–	12	579 581 582
LaSn$_3$		–		–		–		–		–		–		–		–	12	580 581 583
MgAg		–		–		–		–		–		–		–		–	12	585
Mg$_2$Al$_3$		–		–		–	8	1310		–		–		–		–		–
Mg$_3$Al$_2$		–		–		–	8	1310		–		–		–		–		–
Mg$_2$Ge	1	1311		–		–		–		–		–		--		–		–
Mg$_3$Sb$_2$	1	1310		–		–		–		–		–		–		–		–
Mg$_2$Sn	1	1317		–		–	8	1311		–		–		–		–	12	588
MnHg		–		–		–		–		–		–		–		–	12	591
MnNi		–		–		–		–		–		–		–		–	12	592
MnPd		–		–		–		–		–		–		–		–	12	593
MnPt		–		–		–		–		–		–		–		–	12	595
Mn$_3$Pt		–		–		–		–		–		–		–		–	12	594
NaTl		–		–		–		–		–		–		–		–	12	622
NbAl$_3$		–		–		–	8	1322		–		–		–		–		–
Nb$_3$Sn		–		–		–		–		–		–		–		–	12	601
NiAl		–		–	8	1316 1318	8	1321		–		–		–		–		–
Ni$_3$Al		–		–	8	1316 1318	8	1321		–		–		–		–		–
NiSb	1	1327		–		–		–		–		–		–		–		–
Ni$_3$Nb		–		–		–		–		–		–		–		–	12	596
NiTi		–		–		–		–		–		–		–		–	12	597
Ni$_5$Y		–		–		–		–		–		–		–		–	12	598
PbLi		–		–		–		–		–		–		–		–	12	584
PdIn		–		–		–	8	1352		–		–		–		–		–
Pd$_3$Sn		–		–		–		–		–		–		–		–	12	605 607 608
Pd$_3$Yb		–		–		–		–		–		–		–		–	12	606 607 609
PrSn$_3$		–		–		–		–		–		–		–		–	12	611 612 614

Substance Name	Thermal Conductivity		Specif. Heat		Thermal Radiative Properties								Thermal Diffusivity		Viscosity		Thermal Expansion	
					Emissivity		Reflectivity		Absorptivity		Transmissiv.							
	V.	Page	V.	Page	V.	Page	V.	Page	V.	Page	V.	Page	V.	Page	V.	Page	V.	Page
Intermetallic compounds: (continued)																		
PrRu$_2$	–		–		–		–		–		–		–		–		12	610 612 613
RhGe	1	1331	–		–		–		–		–		–		–		–	
RhGe$_2$	1	1331	–		–		–		–		–		–		–		–	
SbGa	–		–		–		–		–		–		–		–		12	450
SbIn	–		–		–		–		–		–		–		–		12	455
SbLa	–		–		–		–		–		–		–		–		12	460
SmAg$_3$	–		–		–		–		–		–		–		–		12	615
Sr$_2$Sn	1	1344	–		–		–		–		–		–		–		–	
TaGe$_2$	1	1348	–		–		–		–		–		–		–		–	
TiAl	–		–		8	1338 1339	8	1341	–		–		–		–		–	
TiCr$_2$	–		–		8	1343 1344	8	1346	–		–		–		–		–	
TiNi	1	1361	–		–		–		–		–		–		–		–	
Ti$_2$Pb	1	1349	–		–		–		–		–		–		–		–	
YbZn$_2$	–		–		–		–		–		–		–		–		12	625
Zn$_2$Zr	–		–		–		–		–		–		–		–		12	628
Invar, super	–		–		–		–		–		–		–		–		12	1182 1183
Iodide hafnium	–		–		7	280	–		–		–		–		–		–	
Iodide titanium	–		4	257	–		–		–		–		10	196	–		–	
Iodide zirconium	–		4	268	–		–		–		–		–		–		–	
Iodine, I$_2$	2	83	6s	53	–		8	78	–		–		10	80	–		–	
Iodine, monatomic	–		6s	54	–		–		–		–		–		11	35	–	
Iodine bromide	–		6s	54	–		–		–		–		–		–		–	
Iodine chloride	–		6s	54	–		–		–		–		–		–		–	
Iodine fluoride	–		6s	54	–		–		–		–		–		–		–	
Iodine heptafluoride	–		6s	55	–		–		–		–		–		–		–	
Iodine pentafluoride	–		6s	55	–		–		–		–		–		–		–	
Iodobenzene	–		6s	55	–		–		–		–		–		–		–	
Iodofluoride	–		6s	54	–		–		–		–		–		–		–	
Iodomethane	–		6s	55	–		–		–		–		–		–		–	
1-Iodo-3-methylbutane	–		6s	55	–		–		–		–		–		–		–	
Iodyride, AgI	2	563	–		–		–		–		–		–		–		–	
Ionium	1	381	–		–		–		–		–		–		–		–	
Iridium, Ir	1	152	4	99	7	289 291	7	294 297 299	–		–		10	81	–		12	153
Iridium alloys:																		
Ir + Os	–		–		–		–		–		–		–		–		12	836
Ir + Pt	–		–		–		–		–		–		–		–		12	836
Ir + Re	–		–		–		–		–		–		–		–		12	836

| Substance Name | Thermal Conductivity | | Specif. Heat | | Thermal Radiative Properties | | | | | | | | | | | | Thermal Diffusivity | | Viscosity | | Thermal Expansion | |
| | | | | | Emissivity | | Reflectivity | | Absorptivity | | Transmissiv. | | | | | | | | | | | |
	V.	Page	V.	Page	V.	Page	V.	Page	V.	Page	V.	Page	V.	Page	V.	Page	V.	Page
Iridium alloys: (continued)																		
Ir + Rh		–		–		–		–		–		–		–		–	12	836
Ir + Os + ΣXi		–		–		–		–		–		–		–		–	12	1122 1126 1127
Ir + Pt + ΣXi		–		–		–		–		–		–		–		–	12	1123 1126 1128
Ir + Re + ΣXi		–		–		–		–		–		–		–		–	12	1124 1126 1129
Ir + Rh + ΣXi		–		–		–		–		–		–		–		–	12	1125 1126 1130
Iridium oxide, IrO_2		–		–		–		–		–		–		–		–	13	270
Iron, Fe	1	156	4	102	7	302 306 310 316	7	319 321 324	7	327 329 332		–	10	82		–	12	157
Iron, Armco	1	157 158 159 160 161 163	4	102	7	303 308	7	322	7	332		–	10	84 95		–	12	160 163 164
Iron, Armco, oxidized		–		–	9	1297	9	1299		–		–		–		–		–
Iron, Armco 21-6-9		–		–		–		–		–		–		–		–	12	1148
Iron, cast	1	1129 1130 1133 1134 1136 1137 1205 1222		–		–		–		–		–		–		–	12	1131 1134
Iron, cast, black temper	1	1137		–		–		–		–		–		–		–		–
Iron, cast, gray hot mold	1	1135		–		–		–		–		–		–		–		–
Iron, cast, heat resistant	1	1146		–		–		–		–		–		–		–		–
Iron, cast, high duty	1	1133 1135		–		–		–		–		–		–		–		–
Iron, cast, nickel-resist	1	1204		–		–		–		–		–		–		–		–
Iron, cast, Nr 1510, spherical	1	1222		–		–		–		–		–		–		–		–
Iron, cast, Nr 1520, pearlitic matrix	1	1222		–		–		–		–		–		–		–		–
Iron, cast, white	1	1130 1135		–		–		–		–		–		–		–		–
Iron, cast, white temper	1	1137		–		–		–		–		–		–		–		–
Iron, electrolytic	1	157 159 160 161	4	103 104		–		–		–		–		–		–	12	160 162 164
Iron, electromagnetic		–		–		–		–		–		–	10	94		–		–
Iron, galvanized		–		–		–	9	781		–		–		–		–		–
Iron, gray cast	1	1130 1135		–		–		–		–		–		–		–		–
Iron, gray cast hot mold	1	1135		–		–		–		–		–		–		–		–
Iron, gray soft cast	1	1135		–		–		–		–		–		–		–		–

Substance Name	Thermal Conductivity		Specif. Heat		Thermal Radiative Properties								Thermal Diffusivity		Viscosity		Thermal Expansion	
					Emissivity		Reflectivity		Absorptivity		Transmissiv.							
	V.	Page	V.	Page	V.	Page	V.	Page	V.	Page	V.	Page	V.	Page	V.	Page	V.	Page
Iron, hiperm	–		–		–		–		–		–		10	94	–		–	
Iron, ingot	1	1134	–		–		–		–		–		–		–		–	
Iron, Nodular	1	1137 1222	–		–		–		–		–		–		–		–	
Iron, oxidized	–		–		9	1296	9	1298	–		–		–		–		–	
Iron, Russian, pearlitic pig	1	1137	–		–		–		–		–		–		–		–	
Iron, silal	1	1222 1223	–		–		–		–		–		–		–		–	
Iron, Swedish	1	158	–		–		–		–		–		–		–		–	
Iron, wrought	1	1185 1219	–		–		–		–		–		–		–		–	
Iron, wrought Duco No. 71 black	–		–		–		9	498	–		–		–		–		–	
Iron aluminide + iron disilicide, cermet	–		–		–		–		–		–		10	529	–		–	
Iron aluminum oxide, $FeO \cdot Al_2O_3$	–		5	1443	–		–		–		–		–		–		–	
Iron boride, Fe_2B	–		–		–		–		–		–		–		–		13	796
Iron calcium magnesium silicates:																		
$FeCaMg_2Si_4O_{12}$	–		–		–		–		–		–		–		–		13	710
$Fe_2CaMgSi_2O_8$	–		–		–		–		–		–		–		–		13	710
Iron calcium silicate, $FeCaSiO_4$	–		–		–		–		–		–		–		–		13	710
Iron carbide, Fe_3C	2	578	5	424	–		–		–		–		–		–		–	
Iron carbonate, $FeCO_3$	–		–		–		–		–		–		–		–		13	642
Iron chlorides:																		
$FeCl_2$	–		5	832	–		–		–		–		–		–		–	
$FeCl_3$	–		–		–		–		–		–		–		–		13	1013
Iron chromite	–		5	1443	–		–		–		–		–		–		–	
Iron chromium oxide, $FeO \cdot Cr_2O_3$	–		5	1446	–		–		–		–		–		–		–	
Iron cobalt oxide, $FeO \cdot Co_2O_3$	–		5	1449	–		–		–		–		–		–		–	
Iron cobaltite	–		5	1449	–		–		–		–		–		–		–	
Iron disilicide + cobalt disilicide, mixture	–		–		–		–		–		–		10	527	–		–	
Iron fluoride, FeF_2	–		5	940	–		–		–		–		–		–		13	1030
Iron-lutetium intermetallic compound, $Fe_{17}Lu_2$	–		–		–		–		–		–		–		–		12	571
Iron-nickel intermetallic compound, $FeNi_3$	–		–		–		–		–		–		–		–		12	574
Iron oxides:																		
FeO	–		5	107	–		–		–		–		–		–		13	271
Fe_2O_3	–		5	110	8	280 282	8	284	–		–		10	391	–		13	274
Fe_3O_4	2	154	5	114	8	282	–		–		–		–		–		13	278
Fe_3O_4, magnetic	2	154	–		–		–		–		–		–		–		–	
Iron oxide + magnesium oxide + ΣXi, mixture	2	483	–		–		–		–		–		–		–		–	
Iron oxide + silicon oxide, mixture	2	410	–		–		–		–		–		–		–		–	
Iron phosphide, Fe_3P	–		–		–		–		–		–		–		–		13	1183
Iron-rhodium intermetallic compound, $FeRh$	–		–		–		–		–		–		–		–		12	575

Substance Name	Thermal Conductivity V.	Page	Specif. Heat V.	Page	Emissivity V.	Page	Reflectivity V.	Page	Absorptivity V.	Page	Transmissiv. V.	Page	Thermal Diffusivity V.	Page	Viscosity V.	Page	Thermal Expansion V.	Page
Iron selenides:																		
FeSe$_2$		–	5	527		–		–		–		–		–		–		–
Fe$_3$Se$_4$		–	5	536		–		–		–		–		–		–		–
Fe$_7$Se$_3$		–	5	533		–		–		–		–		–		–		–
Nonstoichiometric		–	5	530		–		–		–		–		–		–		–
Iron silicate Fe$_2$SiO$_4$		–	5	1452		–		–		–		–	10	416		–	13	710
Iron silicate + magnesium silicate, mixture		–		–		–		–		–		–	10	427		–		–
Iron silicides:																		
FeSi		–	5	577		–		–		–		–	10	468		–	13	1212
FeSi$_2$		–		–		–		–		–		–		–		–	13	1212
Fe$_3$Si		–	5	583		–		–		–		–		–		–	13	1212
Fe$_5$Si$_3$		–	5	580		–		–		–		–		–		–		–
Iron sulfate heptahydrate, FeSO$_4$·7H$_2$O		–	5	1200		–		–		–		–		–		–		–
Iron sulfides:																		
FeS		–	5	674		–		–		–		–		–		–		–
FeS$_2$		–	5	677		–		–		–		–		–		–		–
Nonstoichiometric		–	5	671		–		–		–		–		–		–		–
Iron telluride, FeTe$_2$		–	5	729		–		–		–		–		–		–		–
Iron telluride, nonstoichiometric		–	5	726		–		–		–		–		–		–		–
Iron titanium oxide, FeO·TiO$_2$		–	5	1455		–		–		–		–		–		–		–
Iron vitriol		–	5	1200		–		–		–		–		–		–		–
Iron-yttrium intermetallic compound, Fe$_{17}$Y$_2$		–		–		–		–		–		–		–		–	12	576
Irtran 1, MgF$_2$		–		–	8	951	8	953	8	956	8	960		–		–	13	1045
Irtran 2, ZnS		–		–	8	1214 1216	8	1223	8	1225	8	1228		–		–		–
Irtran 3, CaF$_2$		–		–		–		–	8	930	8	933		–		–	13	1027
Irtran 4, ZnSe		–		–	8	1113 1115	8	1119	8	1122	8	1125		–		–		–
Irtran 5, MgO		–		–	8	296		–		–	8	325		–		–	13	291
Irtran 6, CdTe		–		–	8	1240	8	1242		–	8	1247		–		–	13	1245
Isoamyl acetate		–	6s	56		–		–		–		–		–		–		–
Isoamyl alcohol		–	6s	61		–		–		–		–		–		–		–
pri-Isoamyl alcohol		–	6s	61		–		–		–		–		–		–		–
Isoamyl bromide		–	6s	5		–		–		–		–		–		–		–
Isoamyl iodide		–	6s	55		–		–		–		–		–		–		–
β-Isoamylene		–	6s	61		–		–		–		–		–		–		–
Isotron 22	2	197	6	218		–		–		–		–		–		–		–
Isotron 113	3	201	6	224		–		–		–		–		–		–		–
Isotron 114	3	205	6 6s	228 29		–		–		–		–		–		–		–
Ivory	2	1076		–		–		–		–		–		–		–		–
Ivory, african	2	1076		–		–		–		–		–		–		–		–
Kaolin		–		–	8	1653	8	618 1665		–		–		–		–		–

Substance Name	Thermal Conductivity		Specif. Heat		Thermal Radiative Properties								Thermal Diffusivity		Viscosity		Thermal Expansion	
					Emissivity		Reflectivity		Absorptivity		Transmissiv.							
	V.	Page	V.	Page	V.	Page	V.	Page	V.	Page	V.	Page	V.	Page	V.	Page	V.	Page
Kaolin fibers	–		–		–		–		–		–		10	573	–		–	
Kapok	2	1077	–		–		–		–		–		–		–		–	
Kapton	–		–		–		–		8	1714	8	1716	–		–		–	
Ketene	–		6s	57	–		–		–		–		–		–		–	
Ketopropane	3	129	–		–		–		–		–		–		–		–	
Kieselguhr earth	2	814	–		–		–		–		–		–		–		–	
Kieselguhr earth, ignited	2	814	–		–		–		–		–		–		–		–	
Kieselguhr earth, ordinary	2	814	–		–		–		–		–		–		–		–	
Kogasin I, dodecane	–		6s	34	–		–		–		–		–		–		–	
Kogasin II, tetracane	–		6s	90	–		–		–		–		–		–		–	
Koldboard	2	1125	–		–		–		–		–		–		–		–	
Krypton, Kr	3	50	6	34	–		–		–		–		–		11	37	13	139
Krypton-neon, mixture	3	284	–		–		–		–		–		–		11	279	–	
Krypton-nitrogen, mixture	3	354	–		–		–		–		–		–		–		–	
Krypton-oxygen, mixture	3	356	–		–		–		–		–		–		–		–	
Krypton-xenon, mixture	3	288	–		–		–		–		–		–		11	281	–	
Lacquer, Illinois bronze white	–		–		–		9	518	–		–		–		–		–	
Lacquer, Kemaoryl black	–		–		9	81	9	86	9	89	–		–		–		–	
Lacquer, Kemaoryl white	–		–		9	290	–		9	293 295	–		–		–		–	
Lacquer, Kodak black brushing	–		–		9	521	9	522	–		–		–		–		–	
Lacquer, No. 519 Duro-Lac black	–		–		–		9	541	–		–		–		–		–	
Laminac 4129	–		–		–		–		–		–		–		–		13	1512
Lampblack	2	6	–		–		–		–		–		–		–		–	
Lampblack, SA-25 graphitized	–		5	9	–		–		–		–		–		–		–	
Lanthana	–		5	118	–		–		–		–		–		–		–	
Lanthanum, La	1	171	4	110	–		–		–		–		10	101	–		12	173
Lanthanum alloy, La + Nd + ΣXi	1	988	–		–		–		–		–		–		–		–	
Lanthanum antimonide, LaSb	–		–		–		8	1352	–		–		–		–		–	
Lanthanum boride, LaB_6	–		–		8	723	–		–		8	727	–		–		–	
Lanthanum carbides:																		
LaC_2	–		–		–		–		–		–		–		–		13	935
La_2C_3	–		–		–		–		–		–		–		–		13	935
Lanthanum fluoride, LaF_3	2	633	–		–		–		–		8	994	–		–		13	1032
Lanthanum oxide, La_2O_3	–		5	118	–		8	546	–		–		–		–		13	282
Lanthanum-ruthenium intermetallic compound, $LaRu_2$	–		–		–		–		–		–		–		–		12	579 581 582
Lanthanum selenides:																		
LaSe	1	1301	–		–		–		–		–		–		–		–	
La_2Se_3	–		–		–		–		–		–		–		–		13	1192
Lanthanum silicide, $LaSi_2$	–		–		–		–		–		–		–		–		13	1212

| Substance Name | Thermal Conductivity | | Specif. Heat | | Thermal Radiative Properties | | | | | | | | Thermal Diffusivity | | Viscosity | | Thermal Expansion | |
| | | | | | Emissivity | | Reflectivity | | Absorptivity | | Transmissiv. | | | | | | | |
	V.	Page	V.	Page	V.	Page	V.	Page	V.	Page	V.	Page	V.	Page	V.	Page	V.	Page
Lanthanum sulfides:																		
LaS	2	702	–		–		–		–		–			–		–	13	1239
La$_2$S$_3$	–		–		8	1232	–		–		–			–		–	13	1239
Lanthanum telluride, LaTe	1	1304	–		–		–		–		–			–		–	13	1270
Lanthanum-tin intermetallic compound, LaSn$_3$	–		–		–		–		–		–			–		–	12	580 581 583
Lead, Pb	1	175	4	113	7	335 337	–		7	339 341 343 345	–		10	102	–		12	178
Lead, pyrometric standard	1	183 184	–		–		–		–		–			–		–		–	
Lead alloys:																		
Pb + Ag	1	646	–		–		–		–		–			–		–		–	
Pb + Bi	1	640	–		–		–		–		–			–		–	12	681
Pb + Cd	–		–		–		–		–		–			–		–	12	689
Pb + In	1	643	–		–		–		–		–			–		–	12	827
Pb + Sb	1	637	–		–		–		–		–			–		–		–	
Pb + Sn	1	652	4	363	7	948	–		–		–			–		–	12	872
Pb + Sn, solder	–		4	446	–		–		–		–			–		–		–	
Pb + Tl	1	649	–		–		–		–		–			–		–	12	876
Pb + Sb + ΣXi	1	991	–		–		–		–		–			–		–		–	
Pb + Sb + ΣXi, SAE bearing alloy 12	1	991	–		–		–		–		–			–		–		–	
Lead aluminum oxide, PbO·Al$_2$O$_3$	–		–		–		–		–		–			–		–	13	478
Lead boron silicate, Pb$_5$B$_2$SiO$_{10}$	–		–		–		–		–		–			–		–	13	711
Lead bromide, PbBr$_2$	–		–		–		–		–		8	745		–		–	13	806
Lead carbonate	–		–		–		8	587	8	589	–			–		–		–	
Lead chloride, PbCl$_2$	–		–		–		–		–		8	908		–		–	13	977
Lead chloroiodide, PbClI	–		–		–		–		–		–			–		–	13	1122
Lead fluoride, PbF$_2$	–		–		–		–		–		8	994		–		–	13	1034
Lead fluorochloride, PbFCl	–		–		–		–		–		–			–		–	13	1013
Lead germanium oxide, 2PbO·GeO$_2$	–		–		–		–		–		–			–		–	13	491
Lead germanium telluride, PbGeTe	–		–		–		–		–		–			–		–	13	1271
Lead glance	–		5	681	–		–		–		–			–		–		–	
Lead hafnate	–		–		–		8	597	–		–			–		–		–	
Lead iodide, PbI$_2$	–		5	497	–		–		–		8	1003		–		–	13	1122
Lead iron tungsten oxide, 3PbO·Fe$_2$O$_3$·WO$_3$	–		–		–		–		–		–			–		–	13	590
Lead-lithium intermetallic compound, PbLi	–		–		–		–		–		–			–		–	12	584
Lead molybdenum oxide, PbO·MoO$_3$	–		5	1458	–		–		–		–			–		–		–	
Lead nitrate, Pb(NO$_3$)$_2$	–		–		–		–		–		–			–		–	13	671
Lead oxide + silicon dioxide, mixture	2	359	–		–		–		–		–			–		–		–	
Lead oxide + silicon dioxide + ΣXi, mixture	2	474	–		–		–		–		–			–		–		–	

Substance Name	Thermal Conductivity		Specif. Heat		Thermal Radiative Properties								Thermal Diffusivity		Viscosity		Thermal Expansion	
					Emissivity		Reflectivity		Absorptivity		Transmissiv.							
	V.	Page	V.	Page	V.	Page	V.	Page	V.	Page	V.	Page	V.	Page	V.	Page	V.	Page
Lead oxides:																		
Litharge		–		–		–		–		–		–	10	392		–		–
PbO		–	5	122		–	8	546		–		–	10	392		–		–
PbO₂		–	5	125		–		–		–		–		–		–		–
Pb₂O₃		–	5	128		–		–		–		–		–		–		–
Pb₃O₄		–	5	131		–		–		–		–		–		–		–
Lead phosphates:																		
PbP₂O₆		–		–		–		–		–		–		–		–	13	674
Pb₂P₂O₇		–		–		–		–		–		–		–		–	13	674
Pb₃P₂O₈		–		–		–		–		–		–		–		–	13	674
Pb₃P₄O₁₃		–		–		–		–		–		–		–		–	13	674
Pb₅P₄O₁₅		–		–		–		–		–		–		–		–	13	674
Lead selenide		–		–		–	8	1130		–	8	1133		–		–		–
Lead silicates:																		
PbSiO₃		–		–		–		–		–		–		–		–	13	711
Pb₂SiO₄		–		–		–		–		–		–		–		–	13	711
Pb₄SiO₆		–		–		–		–		–		–		–		–	13	711
Lead sulfate		–		–		–		–		–	8	631		–		–		–
Lead sulfide, PbS		–	5	681		–	8	1197		–	8	1204		–		–		–
Lead sulfite, PbSO₃		–		–		–		–		–	8	631		–		–		–
Lead tantalum oxides:																		
PbO·Ta₂O₅		–		–		–		–		–		–		–		–	13	535
2PbO·Ta₂O₅		–		–		–		–		–		–		–		–	13	537
3PbO·2Ta₂O₅		–		–		–		–		–		–		–		–	13	538
Lead telluride, PbTe	1	1307		–		–	8	1253		–	8	1255	10	470		–	13	1250
Lead + tin, liquid mixture		–		–		–		–		–		–		–	11	576		–
Lead titanium oxide, PbO·TiO₂	2	279		–		–	8	646 648		–	8	650		–		–	13	564
Lead tungsten oxide, PbO·WO₃		–	5	1461		–	8	666		–		–		–		–	13	587
Lead zirconium oxide, PbO·ZrO₂	2	282		–		–	8	676		–		–		–		–	13	606
Lexan		–		–		–		–		–		–		–		–	13	1405
Lignum vitae	2	1079		–		–		–		–		–		–		–		–
Limestone, Indiana	2	821		–		–		–		–		–		–		–		–
Limestone, Queenstone grey	2	821		–		–		–		–		–		–		–		–
Limestone, Rama	2	821		–		–		–		–		–		–		–		–
Limonite, mineral Fe₂O₃		–		–		–	8	1678		–		–		–		–		–
Litharge, lead oxide		–		–		–		–		–		–	10	392		–		–
Lithia	2	157	5	134		–		–		–		–		–		–		–
Lithium, Li	1	192	4	117		–		–		–		–	10	107		–	12	186
Lithium alloys:																		
Li + Mg		–	4	366		–		–		–		–		–		–		–
Li + B + ΣXᵢ	1	992		–		–		–		–		–		–		–		–
Lithium aluminate + strontium oxide + ΣXᵢ, mixture	2	513		–		–		–		–		–		–		–		–

Substance Name	Thermal Conduc- tivity		Specif. Heat		Thermal Radiative Properties								Thermal Diffu- sivity		Visco- sity		Thermal Expan- sion	
					Emis- sivity		Reflec- tivity		Absorp- tivity		Trans- missiv.							
	V.	Page	V.	Page	V.	Page	V.	Page	V.	Page	V.	Page	V.	Page	V.	Page	V.	Page
Lithium aluminum fluoride, Li_3AlF_6		–	5	947		–		–		–		–		–		–		–
Lithium aluminum oxide, $LiAlO_2$		–	5	1464		–		–		–		–		–		–		–
Lithium aluminum silicate, $LiAlSiO_4$		–		–		–		–		–		–		–		–	13	713
Lithium beryllium fluoride, Li_2BeF_4		–	5	950		–		–		–		–		–		–		–
Lithium borate		–		–		–	8	582		–		–		–		–		–
Lithium bromide, LiBr		–		–		–		–		–		–		–		–	13	836
Lithium carbonate, Li_2CO_3		–	5	1118		–		–		–		–		–		–		–
Lithium chloride, LiCl		–	5	835		–		–		–		–		–		–	13	979
Lithium deuteride, LiD		–		–		–		–		–		–		–		–	13	1079
Lithium fluoride, LiF	2	636	5	943		–	8	937 942		–	8	944	10	471		–	13	1036
Lithium fluoride + nickel fluoride, mixture		–		–		–		–		–		–		–		–	13	1075
Lithium fluoride + potassium fluoride + ΣXi, mixture	2	641		–		–		–		–		–		–		–		–
Lithium germanium oxides:																		
$Li_2O \cdot GeO_2$		–		–		–		–		–		–		–		–	13	494
$Li_2O \cdot 7GeO_2$		–		–		–		–		–		–		–		–	13	494
$2Li_2O \cdot GeO_2$		–		–		–		–		–		–		–		–	13	494
$3Li_2O \cdot 2GeO_2$		–		–		–		–		–		–		–		–	13	494
$3Li_2O \cdot 8GeO_2$		–		–		–		–		–		–		–		–	13	494
Lithium hexafluoroaluminate		–	5	947		–		–		–		–		–		–		–
Lithium hydride, LiH	2	773	5	1036		–		–		–		–		–		–	13	1079
Lithium hydrogen fluoride, $LiHF_2$		–	5	953		–		–		–		–		–		–		–
Lithium hydrozinium sulfate, $Li(N_2H_5)SO_4$		–		–		–		–		–		–		–		–	13	734
Lithium iron oxide, $Li_2O \cdot Fe_2O_3$		–	5	1467		–		–		–		–		–		–		–
Lithium iron oxide, nonstoichiometric		–	5	1470		–		–		–		–		–		–		–
Lithium niobium oxide, $Li_2O \cdot Nb_2O_5$		–		–		–	8	598		–		–		–		–	13	526
Lithium oxide, Li_2O	2	157	5	134		–		–		–		–		–		–		–
Lithium silicates:																		
$Li_2Si_2O_5$		–		–		–		–		–		–		–		–	13	713
Li_4SiO_4		–		–		–		–		–		–		–		–	13	713
Li + Na	1	655		–		–		–		–		–		–		–		–
Li + Na + ΣXi	1	995		–		–		–		–		–	10	292		–		–
Lithium sulfate, Li_2SO_4		–		–		–		–		–		–		–		–	13	731
Lithium tantalum oxide, $Li_2O \cdot Ta_2O_5$		–		–		–		–		–		–		–		–	13	539
Lithium titanium oxide, $Li_2O \cdot TiO_2$		–	5	1473		–		–		–		–		–		–		–
Lithium yttrium fluoride		–		–		–		–		–	8	994		–		–		–
Lithium zinc iron oxide, nonstoichiometric		–	5	1476		–		–		–		–		–		–		–
Lithium zirconium silicate + strontium oxide + ΣXi, mixture	2	514		–		–		–		–		–		–		–		–
Lithopone		–		–		–		–		–		–	10	520		–		–
Lucalox	2	106		–		–		–		–		–	10	383		–		–
Lucite		–		–		–	8	1720	8	1722	8	1724		–		–		–

Substance Name	Thermal Conductivity		Specif. Heat		Thermal Radiative Properties								Thermal Diffusivity		Viscosity		Thermal Expansion	
					Emissivity		Reflectivity		Absorptivity		Transmissiv.							
	V.	Page	V.	Page	V.	Page	V.	Page	V.	Page	V.	Page	V.	Page	V.	Page	V.	Page
Lunar materials		–		–		–	8	1666		–		–		–		–		–
Lutetia		–	5	137		–		–		–		–		–		–		–
Lutetium, Lu	1	198	4	121		–	7	347		–	7	350	10	108		–	12	190
Lutetium boride, LuB$_{12}$		–		–		–		–		–		–		–		–	13	793
Lutetium deuteride, LuD		–		–		–		–		–		–		–		–	13	1083
Lutetium hydride, LuH		–		–		–		–		–		–		–		–	13	1083
Lutetium oxide, Lu$_2$O$_3$		–	5	137	8	286 288		–		–		–		–		–	13	285
Magnesia	2	158	5	140		–		–		–		–		–		–		–
Magnesio-ferrite, MO·Fe$_2$O$_3$		–		–		–		–		–		–		–		–	13	513
Magnesio-wustite, MgO·2FeO		–		–		–		–		–		–		–		–	13	510
Magnesium, Mg	1	202	4	124	7	353	7	356 358 360	7	364	7	367	10	109		–	12	194
Magnesium, anodized		–		–	9	1274		–	9	1275		–		–		–		–
Magnesium, L120		–		–		–	7	361 362	7	365		–		–		–		–
Magnesium alloys:																		
Mg + Ag	1	678		–		–		–		–		–		–		–	12	881
Mg + Al	1	658		–		–	7	950		–		–		–		–	12	646
Mg + Al, magnox A-12		–		–		–		–		–		–		–		–	12	649
Mg + Al, magnox C		–		–		–		–		–		–		–		–	12	649
Mg + Ca	1	662	4	294		–		–		–		–		–		–		–
Mg + Ce	1	663		–		–		–		–		–		–		–		–
Mg + Cd	1	661	4	297		–		–		–		–		–		–	12	693
Mg + Cu	1	666		–		–		–		–		–		–		–	12	767
Mg + Mn	1	669		–		–		–		–		–		–		–	12	878
Mg + Ni	1	672		–		–		–		–		–		–		–	12	877
Mg + Si	1	675	4	369		–		–		–		–		–		–		–
Mg + Sn	1	679		–		–		–		–		–		–		–	12	884
Mg + Zn	1	680		–		–		–		–		–		–		–	12	888
Mg + Al + ΣXi	1	998	4	535	7	1327	7	1330	7	1334		–	10	293		–	12	1202
Mg + Al + ΣXi, anodized		–		–		–	9	1277		–		–		–		–		–
Mg + Al + ΣXi, AN-M-29	1	999	4	535		–		–		–		–	10	294		–	12	1204
Mg + Al + ΣXi, AZ31		–		–	7	1328	7	1332	7	1334		–		–		–		–
Mg + Al + ΣXi, AZ31A	1	999		–		–		–		–		–		–		–	12	1202 1204
Mg + Al + ΣXi, AZ31B		–	4	535		–	7	1332		–		–		–		–		–
Mg + Al + ΣXi, AZ31B, anodized		–		–		–	9	1277		–		–		–		–		–
Mg + Al + ΣXi, AZ-80		–	4	535		–		–		–		–		–		–		–
Mg + Al + ΣXi, elckton 2	1	999		–		–		–		–		–		–		–		–
Mg + Ce + ΣXi	1	1001		–		–		–		–		–		–		–		–
Mg + Co + ΣXi	1	1004		–		–		–		–		–		–		–		–
Mg + Cu + ΣXi	1	1005		–		–		–		–		–		–		–	12	1207 1210 1211

Substance Name	Thermal Conductivity		Specif. Heat		Emissivity		Reflectivity		Absorptivity		Transmissiv.		Thermal Diffusivity		Viscosity		Thermal Expansion	
	V.	Page	V.	Page	V.	Page	V.	Page	V.	Page	V.	Page	V.	Page	V.	Page	V.	Page
Magnesium alloys: (continued)																		
Mg + Ni + ΣXi	1	1008		–		–		–		–		–		–		–		–
Mg + Th + ΣXi		–	4	538	7	1336	7	1338	7	1340		–	10	295		–	12	1208 1210 1212
Mg + Th + ΣXi, anodized		–		–		–	9	1281		–		–		–		–		–
Mg + Th + ΣXi, ASTM B80 HZ-32A		–		–		–		–		–		–		–		–	12	1208 1212
Mg + Th + ΣXi, HK-31		–		–	7	1336	7	1338	7	1340		–	10	296		–		–
Mg + Th + ΣXi, HK-31A		–	4	538		–	9	1281		–		–	10	296		–		–
Mg + Th + ΣXi, HM-21XA		–	4	538		–		–		–		–		–		–		–
Mg + Th + ΣXi, HM-31XA		–	4	538		–		–		–		–		–		–		–
Mg + Th + ΣXi, HZ-32A		–		–		–		–		–		–		–		–	12	1208 1212
Mg + Zn + ΣXi		–	4	541		–		–		–		–		–		–	12	1209 1210 1213
Mg + Zn + ΣXi, ASTM B80 ZH-32A		–		–		–		–		–		–		–		–	12	1213
Mg + Zn + ΣXi, ASTM B90 HM-21A		–		–		–		–		–		–		–		–	12	1212
Mg + Zn + ΣXi, elektron AMT		–		–		–		–		–		–		–		–	12	1209
Mg + Zn + ΣXi, ZK-60A		–	4	541		–		–		–		–		–		–		–
Magnesium aluminate, natural ruby spinel	2	284		–		–		–		–		–		–		–		–
Magnesium aluminate + magnesium oxide, mixture	2	362		–		–		–		–		–		–		–		–
Magnesium aluminate + silicon dioxide, mixture	2	365		–		–		–		–		–		–		–		–
Magnesium aluminate + sodium oxide, mixture	2	368		–		–		–		–		–	10	428		–		–
Magnesium-aluminum intermetallic compounds:																		
Mg_2Al_3		–		–		–	8	1310		–		–		–		–		–
Mg_3Al_2		–		–		–	8	1310		–		–		–		–		–
Magnesium aluminum oxides:																		
$MgO \cdot Al_2O_3$	2	283	5	1479		–	8	576		–	8	577	10	418		–	13	479
$2MgO \cdot 7Al_2O_3$	2	286		–		–		–		–		–		–		–		–
Magnesium aluminum silicate, $Mg_2Al_4Si_5O_{18}$		–	5	1503		–		–		–		–		–		–	13	727
Magnesium antimonide, Mg_3Sb_2	1	1310		–		–		–		–		–		–		–		–
Magnesium-antimony intermetallic compound, Mg_3Sb_2	1	1310				–		–		–		–		–		–		–
Magnesium borides:																		
MgB_2		–	5	345		–		–		–		–		–		–		–
MgB_4		–	5	348		–		–		–		–		–		–		–
Magnesium carbonate, $MgCO_3$	2	776		–		–	8	590		–		–	10	421		–	13	643
Magnesium chlorides:																		
$MgCl_2$		–	5	838		–		–		–		–		–		–		–
$MgCl_2 \cdot H_2O$		–	5	841		–		–		–		–		–		–		–
$MgCl_2 \cdot 2H_2O$		–	5	844		–		–		–		–		–		–		–

Substance Name	Thermal Conductivity		Specif. Heat		Thermal Radiative Properties								Thermal Diffusivity		Viscosity		Thermal Expansion	
					Emissivity		Reflectivity		Absorptivity		Transmissiv.							
	V.	Page	V.	Page	V.	Page	V.	Page	V.	Page	V.	Page	V.	Page	V.	Page	V.	Page
Magnesium chlorides: (continued)																		
MgCl$_2$·4H$_2$O		–	5	847		–		–		–		–		–		–		–
MgCl$_2$·6H$_2$O		–	5	850		–		–		–		–		–		–		–
Magnesium chromite		–	5	1482		–		–		–		–		–		–		–
Magnesium chromium oxide, MgO·Cr$_2$O$_3$		–	5	1482		–		–		–		–		–		–	13	486
Magnesium ferrite		–		–		–		–		–		–		–		–	13	513
Magnesium fluoride, MgF$_2$		–		–		–		–		–		–		–		–	13	1043
Magnesium germanide, Mg$_2$Ge		–	5	481		–		–		–		–	10	374		–		–
Magnesium–germanium intermetallic compound, Mg$_2$Ge	1	1311				–		–		–		–		–		–		–
Magnesium germanium oxide, 2MgO·GeO$_2$		–				–		–		–		–		–		–	13	497
Magnesium iron oxides:																		
MgO·2FeO		–		–		–		–		–		–		–		–	13	510
MgO·Fe$_2$O$_3$		–	5	1485		–		–		–		–		–		–	13	513
Nonstoichiometric		–	5	1488		–		–		–		–		–		–		–
Magnesium lead tungsten oxide, MgO·2PbO·WO$_3$		–		–		–		–		–		–		–		–	13	591
Magnesium molybdenum oxide, MgO·MoO$_3$		–	5	1491		–		–		–		–		–		–		–
Magnesium niobium oxides:																		
MgO·Nb$_2$O$_5$		–		–		–		–		–		–		–		–	13	531
2MgO·Nb$_2$O$_5$		–		–		–		–		–		–		–		–	13	531
3MgO·Nb$_2$O$_5$		–		–		–		–		–		–		–		–	13	531
4MgO·Nb$_2$O$_5$		–		–		–		–		–		–		–		–	13	531
Magnesium nitride, Mg$_3$N$_2$		–	5	1084		–		–		–		–		–		–		–
Magnesium orthosilicate + magnesium oxide, mixture	2	394		–		–		–		–		–		–		–		–
Magnesium oxide, MgO	2	158	5	140	8	290 291 293 295	8	298 299 314	8	319 322	8	323	10	393		–	13	288
Magnesium oxide + magnesium silicate, mixture	2	378		–		–		–		–		–		–		–		–
Magnesium oxide + manganese oxide, mixture	2	398		–		–		–		–		–		–		–		–
Magnesium oxide + nickel oxide, mixture	2	381		–		–		–		–		–		–		–		–
Magnesium oxide + silicon dioxide, magnezit	2	385 481		–		–		–		–		–		–		–		–
Magnesium oxide + silicon dioxide, mixture	2	384		–		–		–		–		–	10	440		–		–
Magnesium oxide + silicon dioxide + ΣXi, mixture	2	484		–		–		–		–		–		–		–		–
Magnesium oxide + talc, mixture	2	550		–		–		–		–		–		–		–		–
Magnesium oxide + tin dioxide, mixture	2	387 416 523		–		–		–		–		–		–		–		–
Magnesium oxide + uranium dioxide, mixture	2	390		–		–		–		–		–		–		–		–
Magnesium oxide + zinc oxide, mixture	2	391 435		–		–		–		–		–		–		–		–

Substance Name	Thermal Conduc- tivity		Specif. Heat		Thermal Radiative Properties									Thermal Diffu- sivity		Visco- sity		Thermal Expan- sion	
					Emis- sivity		Reflec- tivity		Absorp- tivity		Trans- missiv.									
	V.	Page	V.	Page	V.	Page	V.	Page	V.	Page	V.	Page	V.	Page	V.	Page	V.	Page	V.	Page
Magnesium oxide + zirconium oxide, mixture	2	446		–		–		–		–		–	10	451		–		–		
Magnesium phosphate, Mg(PO$_3$)$_2$, mixture		–		–		–	8	608		–		–		–		–	13	690		
Magnesium silicates:																				
MgSiO$_3$		–	5	1497		–	8	618		–		–		–		–	13	715 716 717		
Mg$_2$SiO$_4$	2	275	5	1497		–	8	618		–		–	10	422		–	13	718		
Mg$_3$Si$_4$O$_{11}$·H$_2$O		–	5	1500		–		–		–		–		–		–		–		
Magnesium silicide, Mg$_2$Si	1	1314		–	8	1173		–		–		–		–		–	13	1212		
Magnesium-silver intermetallic compound, MgAg		–		–		–		–		–		–		–		–	12	585		
Magnesium stannate, MgSnO$_3$	2	289		–		–		–		–		–		–		–		–		
Magnesium stannide, Mg$_2$Sn	1	1317		–		–		–		–		–	10	375		–		–		
Magnesium-tin intermetallic compound, Mg$_2$Sn		–		–		–		–		–		–		–		–	12	588		
Magnesium titanium oxides:																				
MgO·TiO$_2$		–	5	1506		–		–		–		–		–		–		–		
MgO·2TiO$_2$		–	5	1509		–		–		–		–		–		–	13	567		
2MgO·TiO$_2$		–	5	1512		–		–		–		–		–		–	13	568		
Magnesium tungsten oxide, MgO·WO$_3$		–	5	1515		–		–		–		–		–		–		–		
Magnesium vanadium oxides:																				
MgO·V$_2$O$_5$		–	5	1518		–		–		–		–		–		–		–		
2MgO·V$_2$O$_5$		–	5	1521		–		–		–		–		–		–		–		
Magnesium zirconium silicate		–		–		–		–	8	616		–		–		–		–		
Mahogany	2	1080		–		–		–		–		–		–		–		–		
Manganese, Mn	1	208	4	127		–		–		–		–	10	111		–	12	201		
Manganese, electrolytic		–	4	127		–		–		–		–		–		–		–		
Manganese alloys:																				
Mn + Al		–	4	372		–		–		–		–		–		–		–		
Mn + Cu	1	683	4	377		–		–		–		–		–		–	12	774		
Mn + Fe	1	684		–		–		–		–		–		–		–	12	842		
Mn + Ni	1	685	4	380		–		–		–		–		–		–	12	892		
Mn + Fe + ΣXi	1	1009		–		–		–		–		–		–		–		–		
Mn + Fe + ΣXi, Russian, ferromanganese	1	684 1010		–		–		–		–		–		–		–		–		
Mn + Fe + ΣXi, Russian, ferromanganese, low carbon	1	1010		–		–		–		–		–		–		–		–		
Mn + Fe + ΣXi, Russian, ferromanganese, normal	1	1010		–		–		–		–		–		–		–		–		
Mn + Fe + ΣXi, Russian, silicomanganese	1	1010 1012		–		–		–		–		–		–		–		–		
Mn + Si + ΣXi	1	1012		–		–		–		–		–		–		–		–		
Manganese aluminum carbide, Mn$_3$AlC		–	5	427		–		–		–		–		–		–		–		
Manganese aluminum oxide, MnO·Al$_2$O$_3$		–		–		–		–		–		–		–		–		–	13	483
Manganese arsenide, MnAs		–				–		–		–		–		–		–	13	752		

Substance Name	Thermal Conductivity		Specif. Heat		Thermal Radiative Properties								Thermal Diffusivity		Viscosity		Thermal Expansion	
					Emissivity		Reflectivity		Absorptivity		Transmissiv.							
	V.	Page	V.	Page	V.	Page	V.	Page	V.	Page	V.	Page	V.	Page	V.	Page	V.	Page
Manganese bromide tetrahydrate, $MnBr_2 \cdot 4H_2O$	–		–		–		–		–		–		–		–		13	808
Manganese carbide, Mn_3C	–		5	433	–		–		–		–		–		–		–	
Manganese carbonate, $MnCO_3$	–		5	1121	–		–		–		–		–		–		13	644
Manganese chlorides:																		
$MnCl_2$	–		5	853	–		–		–		–		–		–		–	
$MnCl_2 \cdot 4H_2O$	–		5	856	–		–		–		–		–		–		13	1013
Manganese fluoride, MnF_2	–		5	959	–		–		–		–		–		–		13	1048
Manganese iron oxide, $MnO \cdot Fe_2O_3$	2	292	–		–		–		–		–		–		–		–	
Manganese–mercury intermetallic compound, MnHg	–				–		–		–		–		–		–		12	591
Manganese–nickel intermetallic compound, MnNi	–				–		–		–		–		–		–		12	592
Manganese oxide + silicon dioxide, mixture	2	.399	–		–		–		–		–		–		–		–	
Manganese oxide + titanium oxide powders	–		–		–		8	563	–		–		–		–		–	
Manganese oxides:																		
MnO	2	168	5	145 151	–		8	329	–		–		–		–		13	302
MnO_2	–		5	148	–		–		–		–		–		–		13	305
Mn_2O_3	–		5	151	–		–		–		–		–		–		13	308
Mn_3O_4	2	170	5	154	–		8	329	–		–		–		–		–	
Manganese–palladium intermetallic compound, MnPd	–		–		–		–		–		–		–		–		12	593
Manganese phosphide, MnP	–		–		–		–		–		–		–		–		13	1172
Manganese–platinum intermetallic compounds:																		
MnPt	–		–		–		–		–		–		–		–		12	595
Mn_3Pt	–		–		–		–		–		–		–		–		12	594
Manganese selenide	–		5	539	–		–		–		–		–		–		–	
Manganese silicate, $MnSiO_3$	–		5	1524	–		–		–		–		–		–		13	727
Manganese silicide, nonstoichiometric	–		5	589	–		–		–		–		–		–		–	
Manganese silicides:																		
MnSi	–		–		–		–		–		–		–		–		13	1212
$MnSi_2$	–		–		8	1173	–		–		–		–		–		–	
Mn_3Si	–		5	586	8	1173	–		–		–		–		–		–	
Manganese sulfide, MnS	–		5	684	–		8	1234	–		–		–		–		13	1225
Manganese telluride, MnTe	–		5	732	–		8	1256	–		–		–		–		13	1253
Manganese zinc carbide, Mn_3ZnC	–		5	430	–		–		–		–		–		–		–	
Manganese zinc ferrate, $MnZnFe_2O_4$	2	295	–		–		–		–		–		–		–		–	
Manganite	–		–		–		–		–		8	1664	–		–		–	
Manganomanganic oxide, Mn_3O_4	2	170	5	154	–		–		–		–		–		–		–	
Manganous chloride tetrahydrate, $MnCl_2 \cdot 4H_2O$	–		5	856	–		–		–		–		–		–		–	
Manganous selenide, MnSe	–		5	539	–		–		–		–		–		–		–	
Manganous telluride, MnTe	–		5	732	–		–		–		–		–		–		–	

Substance Name	Thermal Conductivity		Specif. Heat		Thermal Radiative Properties								Thermal Diffusivity		Viscosity		Thermal Expansion	
					Emissivity		Reflectivity		Absorptivity		Transmissiv.							
	V.	Page	V.	Page	V.	Page	V.	Page	V.	Page	V.	Page	V.	Page	V.	Page	V.	Page
Maple	2	1081		–		–		–		–		–		–		–		–
Marble	2	760 761		–		–		–		–		–	10	547		–		–
Marble, black	2	761		–		–		–		–		–		–		–		–
Marble, brown	2	761		–		–		–		–		–		–		–		–
Marble, brown, calcite	2	761		–		–		–		–		–		–		–		–
Marble, powder	2	760 761		–		–		–		–		–		–		–		–
Marble, white	2	761		–		–		–		–		–		–		–		–
Marble, white, Alabama	2	761		–		–	8	583		–	8	585	10	414		–		–
Marsh gas	3	218	6	244		–		–		–		–		–		–		–
Mercuric oxide, HgO		–	5	157		–		–		–		–		–		–		–
Mercuric selenide, HgSe		–	5	542		–		–		–		–		–		–		–
Mercury, Hg	1	212	4	131		–		–		–		–	10	112		–	12	206
Mercury alloy, Hg + Na	1	686		–		–		–		–		–		–		–		–
Mercury bromide, Hg_2Br_2		–		–		–	8	747		–		–		–		–		–
Mercury chlorides:																		
$\quad$ $HgCl_2$		–		–		–	8	908		–		–		–		–		–
$\quad$ Hg_2Cl_2		–		–		–	8	908		–		–		–		–		–
Mercury iodides:																		
$\quad$ HgI_2		–		–		–	8	1027		–	8	1029		–		–	13	1122
$\quad$ Hg_2I_2		–		–		–	8	1027		–		–		–		–		–
Mercury oxide, HgO		–		–		–		–		–	8	549		–		–		–
Mercury selenide, HgSe	1	1320	5	542		–		–		–		–		–		–	13	1192
Mercury sulfate, Hg_2SO_4		–	5	1203		–		–		–		–		–		–		–
Mercury sulfide, HgS		–	5	687		–		–		–		–		–		–		–
Mercury telluride, HgTe	1	1321		–		–		–		–		–		–		–	13	1256
Mercury telluride + cadmium telluride, mixture	1	1407		–		–		–		–		–		–		–		–
Mesitylene		–	6s	57		–		–		–		–		–		–		–
Metal, rose	1	939		–		–		–		–		–		–		–		–
Methanal		–	6s	42		–		–		–		–		–		–		–
Methane	3	218	6	244		–		–		–		–		–	11	186		–
Methane, dideuterated		–	6s	58		–		–		–		–		–		–		–
Methane, dideuterated ditritiated		–	6s	58		–		–		–		–		–		–		–
Methane, ditritiated		–	6s	58		–		–		–		–		–		–		–
Methane, monodeuterated		–	6s	58		–		–		–		–		–		–		–
Methane, monodeuterated tritritiated		–	6s	58		–		–		–		–		–		–		–
Methane, monotritiated		–	6s	58		–		–		–		–		–		–		–
Methane, tetradeuterated		–	6s	58		–		–		–		–		–		–		–
Methane carboxylic acid		–	6s	1		–		–		–		–		–		–		–
Methane-nitrogen, mixture		–		–		–		–		–		–		–	11	465		–
Methane-oxygen, mixture		–		–		–		–		–		–		–	11	474		–
Methane-propane, mixture	3	432		–		–		–		–		–		–	11	477		–

Substance Name	Thermal Conductivity		Specif. Heat		Thermal Radiative Properties								Thermal Diffusivity		Viscosity		Thermal Expansion	
					Emissivity		Reflectivity		Absorptivity		Transmissiv.							
	V.	Page	V.	Page	V.	Page	V.	Page	V.	Page	V.	Page	V.	Page	V.	Page	V.	Page
Methane-sulfur dioxide, mixture		–		–		–		–		–		–		–	11	529		–
Methanethiol		–	6s	59		–		–		–		–		–		–		–
Methanethiomethane		–	6s	69		–		–		–		–		–		–		–
Methanol	3	223	6	252		–		–		–		–		–		–		–
Methenyl tribromide		–	6s	5		–		–		–		–		–		–		–
Methoxymethane		–	6s	63		–		–		–		–		–		–		–
Methyl		–	6s	59		–		–		–		–		–		–		–
Methyl acetate		–	6s	59		–		–		–		–		–		–		–
Methylacetylene		–	6s	82		–		–		–		–		–		–		–
Methane, tetratritiated		–	6s	58		–		–		–		–		–		–		–
Methane, trideuterated		–	6s	59		–		–		–		–		–		–		–
Methane, trideuterated monotritiated		–	6s	59		–		–		–		–		–		–		–
Methane, tritritiated		–	6s	59		–		–		–		–		–		–		–
Methyl alcohol	3	223	6	252		–		–		–		–		–	11	192		–
Methyl aldehyde		–	6s	42		–		–		–		–		–		–		–
Methyl bromide		–	6s	5		–		–		–		–		–		–		–
Methyl chloride	3	227	6	257		–		–		–		–		–	11	194		–
Methyl chloride-sulfur dioxide, mixture		–		–		–		–		–		–		–	11	551		–
Methyl cyanide		–	6s	61		–		–		–		–		–		–		–
Methyl ethanoate		–	6s	59		–		–		–		–		–		–		–
Methyl ether		–	6s	63		–		–		–		–		–		–		–
Methyl ethyl ketone		–	6s	7		–		–		–		–		–		–		–
Methyl fluoride		–	6s	42		–		–		–		–		–		–		–
Methyl formate-propane, mixture	3	462		–		–		–		–		–		–		–		–
Methyl glycol		–	6s	76		–		–		–		–		–		–		–
Methyl iodide		–	6s	55		–		–		–		–		–		–		–
Methyl isobutyl ketone		–	6s	66		–		–		–		–		–		–		–
Methyl isocyanide		–	6s	64		–		–		–		–		–		–		–
Methyl isonitrile		–	6s	64		–		–		–		–		–		–		–
Methyl mercaptan		–	6s	59		–		–		–		–		–		–		–
Methyl oxide		–	6s	63		–		–		–		–		–		–		–
Methyl sulfide		–	6s	69		–		–		–		–		–		–		–
Methyl thioalcohol		–	6s	59		–		–		–		–		–		–		–
Methylamine		–	6s	59		–		–		–		–		–		–		–
Methylbenzene		–	6	285		–		–		–		–		–		–		–
m-Methylbenzoic acid		–	6s	91		–		–		–		–		–		–		–
o-Methylbenzoic acid		–	6s	91		–		–		–		–		–		–		–
p-Methylbenzoic acid		–	6s	91		–		–		–		–		–		–		–
2-Methylbenzoic acid		–	6s	91		–		–		–		–		–		–		–
3-Methylbenzoic acid		–	6s	91		–		–		–		–		–		–		–
4-Methylbenzoic acid		–	6s	91		–		–		–		–		–		–		–
1,b-Methylbivinyl		–	6s	56		–		–		–		–		–		–		–

Substance Name	Thermal Conductivity		Specif. Heat		Thermal Radiative Properties								Thermal Diffusivity		Viscosity		Thermal Expansion	
					Emissivity		Reflectivity		Absorptivity		Transmissiv.							
	V.	Page	V.	Page	V.	Page	V.	Page	V.	Page	V.	Page	V.	Page	V.	Page	V.	Page
2-Methyl-1,3-butadiene	–		6s	56	–		–		–		–		–		–		–	
3-Methyl-1,3-butadiene	–		6s	56	–		–		–		–		–		–		–	
2-Methylbutane	–		6s	59	–		–		–		–		–		–		–	
2-Methyl-2-butanol	–		6s	61	–		–		–		–		–		–		–	
3-Methyl-1-butanol	–		6s	61	–		–		–		–		–		–		–	
3-Methyl-1-butanol acetate	–		6s	56	–		–		–		–		–		–		–	
2-Methyl-2-butene	–		6s	61	–		–		–		–		–		–		–	
γ-Methylbutyl ethanoate	–		6s	56	–		–		–		–		–		–		–	
3-Methyl-1-butyne	–		6s	61	–		–		–		–		–		–		–	
Methylcarbylamine	–		6s	64	–		–		–		–		–		–		–	
Methylchloroform	–		6s	91	–		–		–		–		–		–		–	
Methylcyclohexane	–		6s	62	–		–		–		–		–		–		–	
Methylcyclopentane	–		6s	62	–		–		–		–		–		–		–	
Methyldipropylmethane	–		6s	64	–		–		–		–		–		–		–	
α-Methylditan	–		6s	34	–		–		–		–		–		–		–	
Methylene	–		6s	63	–		–		–		–		–		–		–	
Methylene bromide	–		6s	26	–		–		–		–		–		–		–	
Methylene chloride	–		6s	28	–		–		–		–		–		–		–	
Methylene dichloride	–		6s	28	–		–		–		–		–		–		–	
Methylene fluoride	–		6s	30	–		–		–		–		–		–		–	
Methylene iodide	–		6s	30	–		–		–		–		–		–		–	
Methylene oxide	–		6s	42	–		–		–		–		–		–		–	
Methylethylene glycol	–		6s	76	–		–		–		–		–		–		–	
sym-Methylethylethylene	–		6s	73	–		–		–		–		–		–		–	
Methylfluoroform	–		6s	92	–		–		–		–		–		–		–	
2-Methylfuran	–		6s	63	–		–		–		–		–		–		–	
2-Methylheptane	–		6s	63	–		–		–		–		–		–		–	
3-Methylheptane	–		6s	64	–		–		–		–		–		–		–	
4-Methylheptane	–		6s	64	–		–		–		–		–		–		–	
2-Methylhexane	–		6s	64	–		–		–		–		–		–		–	
Methylhydrazine	–		6s	64	–		–		–		–		–		–		–	
Methylhydrazine, MMH	–		6s	64	–		–		–		–		–		–		–	
Methylidyne	–		6s	64	–		–		–		–		–		–		–	
Methylmethane	–		6	174	–		–		–		–		–		–		–	
2-Methylpentane	–		6s	64	–		–		–		–		–		–		–	
2-Methyl-3-ethylpentane	–		6s	38	–		–		–		–		–		–		–	
3-Methylpentane	–		6s	65	–		–		–		–		–		–		–	
3-Methyl-3-ethylpentane	–		6s	38	–		–		–		–		–		–		–	
4-Methyl-2-pentanone	–		6s	66	–		–		–		–		–		–		–	
2-Methyl-2-phenylpropane	–		6s	11	–		–		–		–		–		–		–	
2-Methyl-1-propanol	–		6s	67	–		–		–		–		–		–		–	
2-Methyl-2-propanol	–		6s	67	–		–		–		–		–		–		–	

| Substance Name | Thermal Conductivity | | Specif. Heat | | Thermal Radiative Properties | | | | | | | | Thermal Diffusivity | | Viscosity | | Thermal Expansion | |
| | | | | | Emissivity | | Reflectivity | | Absorptivity | | Transmissiv. | | | | | | | |
	V.	Page	V.	Page	V.	Page	V.	Page	V.	Page	V.	Page	V.	Page	V.	Page	V.	Page
2-Methylpropene		–	6s	67		–		–		–		–		–		–		–
β-Methylpropyl acetate		–	6s	55		–		–		–		–		–		–		–
Methylthiomethane		–	6s	69		–		–		–		–		–		–		–
Mica		–		–		–		–		–	8	1694	10	548		–		–
Mica, bonded	2	825		–		–		–		–		–		–		–		–
Mica, Canadian phlogopies	2	824 825		–		–		–		–		–		–		–		–
Mica, Madagascan phlogopites	2	824		–		–		–		–		–		–		–		–
Mica, synthetic	2	825		–		–		–		–		–		–		–		–
Micabond, dull black		–		–	9	530		–	9	531		–		–		–		–
Micanite	2	1138		–		–		–		–		–		–		–		–
Micarta laminates		–		–		–		–		–		–	10	559		–		–
Milk curd		–		–		–		–		–		–	10	641		–		–
Mineral wool, processed felt	2	1141		–		–		–		–		–		–		–		–
Molybdenum, Mo	1	222	4	135	7	376 383 387 392	7	398 402	7	404 407 410		–	10	113		–	12	208
Molybdenum alloys:																		
Mo + Cr		–		–		–		–		–		–		–		–	12	713
Mo + Cu		–		–		–		–		–		–		–		–	12	779
Mo + Fe	1	690		–		–		–		–		–		–		–		–
Mo + Nb		–		–		–		–		–		–		–		–	12	900
Mo + Re		–		–		–		–		–		–		–		–	12	904
Mo + Se		–		–		–		–		–		–		–		–	12	907
Mo + Te		–		–		–		–		–		–		–		–	12	907
Mo + Ti	1	691	4	383	7	953 956 959	7	962		–		–	10	244		–	12	910
Mo + Ti, TZM alloy		–		–		–		–		–		–		–		–	12	912 913
Mo + V		–		–		–		–		–		–		–		–	12	923
Mo + W	1	694	4	386	7	967 969		–		–		–	10	246		–	12	915
Mo + Fe + ΣXi	1	1013		–		–		–		–		–		–		–		–
Mo + Fe + ΣXi, Russian, ferromolybdenum	1	690 1013		–		–		–		–		–		–		–		–
Mo + Ti + ΣXi		–	4	544		–		–		–		–		–		–	12	1214
Molybdenum beryllide, MoBe12		–	5	316		–		–		–		–		–		–		–
Molybdenum borides:																		
MoB		–	5	358	8	692		–		–		–		–		–		–
MoB2		–	5	352		–	8	695		–		–		–		–	13	796
Mo2B		–	5	355		–		–		–		–		–		–		–
MoB5		–		–		–	8	695		–		–		–		–		–

| Substance Name | Thermal Conductivity | | Specif. Heat | | Thermal Radiative Properties | | | | | | | | Thermal Diffusivity | | Viscosity | | Thermal Expansion | |
| | | | | | Emissivity | | Reflectivity | | Absorptivity | | Transmissiv. | | | | | | | |
	V.	Page	V.	Page	V.	Page	V.	Page	V.	Page	V.	Page	V.	Page	V.	Page	V.	Page
Molybdenum carbides:																		
MoC		–	5	436		–		–		–		–		–		–	13	935
Mo_2C	2	579	5	436	8	850 852		–		–		–		–		–	13	854
Molybdenum disilicide, oxidized		–		–		–		–		–	9	1311		–		–		–
Molybdenum disilicide + molybdenum oxide, mixture		–		–	8	1473 1474	8	1476		–		–		–		–		–
Molybdenum disilicide + molybdenum oxide + silicon dioxide powder		–		–	8	1510 1511	8	1513		–		–		–		–		–
Molybdenum disilicide + silicon dioxide powder		–		–	8	1478 1479	8	1482		–		–		–		–		–
Molybdenum fluoride, MoF_6		–	5	962		–		–		–		–		–		–		–
Molybdenum oxide + molybdenum silicide powder		–		–	8	1473 1474	8	1476		–		–		–		–		–
Molybdenum oxide + molybdenum disilicide powder		–		–	8	1510 1511	8	1513		–		–		–		–		–
Molybdenum oxide + nickel, cermet		–		–		–	8	1425		–		–		–		–		–
Molybdenum oxide + silicon dioxide powder		–		–		–	8	566		–		–		–		–		–
Molybdenum oxides:																		
MoO_2		–	5	160		–		–		–		–		–		–		–
MoO_3		–	5	163		–	8	330		–		–		–		–		–
Mo_2O_3		–		–		–		–		–		–		–		–	13	311
Molybdenum silicides:																		
$MoSi_2$	1	1324	5	592	8	1148 1150 1152	8	1155		–		–		–		–	13	1198
$MoSi_2$, oxidized		–		–		–		–		–	9	1311		–		–		–
$MoSi_3$		–		–		–		–		–		–		–		–	13	1212
Mo_3Si		–	5	595	8	1150		–		–		–		–		–	13	1212
Mo_5Si_3		–		–	8	1150	8	1482		–		–		–		–	13	1212
Molybdenum silicide + molybdenum oxide + silicon oxide powder		–		–		–	8	1510 1511	8	1513		–		–		–		–
Molybdenum silicide + silicon oxide, powder		–		–	8	1478 1479		–		–		–		–		–		–
Molybdenum silicide + zirconium carbide + ΣX_i, mixture		–		–		–		–		–		–		–		–	13	790
Molybdenum, siliconized		–		–	9	1331 1333 1335 1337	9	1342	9	1345		–		–		–		–
Molybdenum sulfide, MoS_2		–	5	690		–	8	1207		–	8	1210		–		–	13	1239
Molybdenum telluride, $MoTe_2$		–		–		–		–		–		–	10	473		–		–
Molybdenum telluride + tungsten telluride, mixture		–		–		–		–		–		–	10	531		–		–
Monodeuteriomethane		–	6s	58		–		–		–		–		–		–		–
Mullite, aluminum silicate	2	254 934		–	8	1685 1687	8	618		–		–	10	412		–		–

Substance Name	Thermal Conductivity		Specif. Heat		Thermal Radiative Properties								Thermal Diffusivity		Viscosity		Thermal Expansion	
					Emissivity		Reflectivity		Absorptivity		Transmissiv.							
	V.	Page	V.	Page	V.	Page	V.	Page	V.	Page	V.	Page	V.	Page	V.	Page	V.	Page
Multimet		–		–	7	1214 1227 1238		–		–		–		–		–		–
Multimet, low C		–		–		–		–		–		–		–		–	12	1146
Multimet, med. C		–		–		–		–		–		–		–		–	12	1146
Multimet, N-155	1	1165		–		–		–		–		–		–		–		–
Multiple oxides and salts		–		–		–		–		–		–	10	411		–		–
Mylar		–		–		–		–	8	1708 1710	8	1711		–		–	13	1438
Naphthalene	2	995	6s	69		–		–		–		–		–		–		–
Naphthalin	2	995	6s	69		–		–		–		–		–		–		–
Naphthol	2	998		–		–		–		–		–		–		–		–
α-Naphthol		–	6s	69		–		–		–		–		–		–		–
β-Naphthol		–	6s	69		–		–		–		–		–		–		–
1-Naphthol		–	6s	69		–		–		–		–		–		–		–
2-Naphthol		–	6s	69		–		–		–		–		–		–		–
Natrolite		–		–		–		–		–	8	1694		–		–		–
Neodymia		–	5	166		–		–		–		–		–		–		–
Neodymium, Nd	1	230	4	140		–		–		–		–	10	119		–	12	219
Neodymium boride, NdB_6		–		–	8	723		–		–	8	727		–		–		–
Neodymium carbides:																		
NdC$_2$		–		–		–		–		–		–		–		–	13	935
Nd$_2$C$_3$		–		–		–		–		–		–		–		–	13	936
Neodymium chloride hexahydrate, $NdCl_3 \cdot 6H_2O$		–	5	859		–		–		–		–		–		–		–
Neodymium gallium oxide, $Nd_3Ga_5O_{12}$, garnet		–	5	1527		–		–		–		–		–		–		–
Neodymium oxide, Nd_2O_3		–	5	166		–	8	335		–		–		–		–	13	315
Neodymium selenide, Nd_2Se_3		–		–		–		–		–		–		–		–	13	1192
Neodymium silicate		–		–		–		–		–	8	622		–		–		–
Neodymium sulfides:																		
NdS		–		–		–		–		–		–		–		–	13	1239
Nd$_2$S$_3$		–		–	8	1232		–		–		–		–		–	13	1240
Neohexane		–	6s	31		–		–		–		–		–		–		–
Neon, Ne	3	56	6	37		–		–		–		–		–	11	41	13	144
Neon-nitrogen, mixture	3	365		–		–		–		–		–		–	11	339		–
Neon-nitrogen-oxygen, mixture	3	495		–		–		–		–		–		–		–		–
Neon-oxygen, mixture	3	368		–		–		–		–		–		–		–		–
Neon-xenon, mixture	3	291		–		–		–		–		–		–	11	283		–
Neopentane		–	6s	33		–		–		–		–		–		–		–
Neptunium	1	234	4	143		–		–		·		–		–		–		–
Neptunium alloy, Np + Ca + ΣXi		–	4	547		–		–		–		–		–		–		–
Neptunium nitride, NpN		–		–		–		–		–		–		–		–	13	1162
Neptunium oxide, NpO_2		–	5	169		–		–		–		–		–		–		–
Neptunium phosphide, NpP		–		–		–		–		–		–		–		–	13	1183

| Substance Name | Thermal Conductivity | | Specif. Heat | | Thermal Radiative Properties | | | | | | | | | | Thermal Diffusivity | | Viscosity | | Thermal Expansion | |
| | | | | | Emissivity | | Reflectivity | | Absorptivity | | Transmissiv. | | | | | | | | | |
	V.	Page	V.	Page	V.	Page	V.	Page	V.	Page	V.	Page	V.	Page	V.	Page	V.	Page
Nerofil		–		–		–		–		–		–	10	22		–		–
Nibrobenzene		–	6s	69		–		–		–		–		–		–		–
Nickel, Ni	1	237	4	146	7	413 416 424 434	7	440 446 454 457	7	460 462 465 468		–	10	120		–	12	225
Nickel, 1 percent impurities	1	1044		–		–		–		–		–		–		–		–
Nickel, A	1	239 241 1029 1039		–		–		–		–		–	10	250		–		–
Nickel, D	1	1039		–		–		–		–		–		–		–		–
Nickel, electrolytic	1	238 239 240	4	146		–		–		–		–		–		–	12	227 748
Nickel, L	1	238 239		–		–		–		–		–		–		–	12	227
Nickel, Mond		–	4	146		–		–		–		–		–		–	12	228
Nickel, NP-3		–		–	7	435		–		–		–		–		–		–
Nickel, O	1	239		–		–		–		–		–		–		–		–
Nickel, oxidized		–		–	9	1312		–		–		–		–		–		–
Nickel, S		–		–		–		–		–		–		–		–	12	228
Nickel, spectrographically standardized		–		–		–		–		–		–	10	122		–		–
Nickel alloys:																		
Ni + Ag					7	979		–		–		–		–		–		–
Ni + Al		–	4	389		–		–		–		–		–		–		–
Ni + Co	1	700		–		–		–		–		–		–		–	12	747
Ni + Co, RCA N91	1	701		–		–		–		–		–		–		–		–
Ni + Co, RCA N97	1	701		–		–		–		–		–		–		–		–
Ni + Cr	1	697	4	392	7	972		–		–		–		–		–	12	719
Ni + Cr, Chromel P	1	698	4	392		–		–		–		–		–		–		–
Ni + Cr, Nickrom	1	698		–		–		–		–		–		–		–		–
Ni + Cr, Nichrome N	1	698		–		–		–		–		–		–		–		–
Ni + Cr, Vacromin F	1	1213		–		–		–		–		–		–		–		–
Ni + Cu	1	703	4	398		–		–		–		–		–		–	12	778
Ni + Fe	1	707	4	403	7	976		–		–		–	10	248		–	12	848
Ni + Fe, malleable nickel		–		–		–		–		–		–		–		–	12	852 853
Ni + Fe, N.S. nickel	1	708		–		–		–		–		–		–		–		–
Ni + Fe, Permalloy		–		–		–		–		–		–		–		–	12	853 854
Ni + Mg		–	4	407		–		–		–		–		–		–		–
Ni + Mn	1	710	4	410		–		–		–		–	10	249		–	12	894
Ni + Mo		–		–		–		–		–		–		–		–	12	899
Ni + Pd		–		–		–		–		–		–		–		–	12	926
Ni + Pt		–		–		–		–		–		–		–		–	12	930
Ni + Si		–	4	413		–		–		–		–		–		–	12	932
Ni + Sn		–		–		–		–		–		–		–		–	12	935

Substance Name	Thermal Conductivity		Specif. Heat		Thermal Radiative Properties								Thermal Diffusivity		Viscosity		Thermal Expansion	
					Emissivity		Reflectivity		Absorptivity		Transmissiv.							
	V.	Page	V.	Page	V.	Page	V.	Page	V.	Page	V.	Page	V.	Page	V.	Page	V.	Page
Nickel alloys: (continued)																		
Ni + Ti	–		–		–		–		–		–		–		–		12	938
Ni + V	–		–		–		–		–		–		–		–		12	943
Ni + W	–		4	416	–		–		–		–		–		–		12	941
Ni + Zn	–		4	419	–		–		–		–		–		–		12	946
Ni + Al + ΣXi	1	1014	–		–		–		–		–		–		–		12	1215
Ni + Al + ΣXi, Duranickel	1	1015	–		–		–		–		–		–		–		–	
Ni + Al + ΣXi, Duranickel 301	1	1015	–		–		–		–		–		–		–		–	
Ni + Al + ΣXi, Nickel Z	1	1015	–		–		–		–		–		–		–		–	
Ni + Cr + ΣXi	1	1017	4	550	7	1342 1347 1356 1360	7	1374 1392	7	1395 1398 1400	–		10	298	–		12	1216
Ni + Cr + ΣXi, Astrolloy	–		–		7	1363 1412	7	1377 1418	–		–		–		–		–	
Ni + Cr + ΣXi, Chroman	1	1018	–		–		–		–		–		–		–		–	
Ni + Cr + ΣXi, Chromel A	1	698	4	556	–		–		–		–		–		–		–	
Ni + Cr + ΣXi, Chromel C	1	1036	–		–		–		–		–		–		–		–	
Ni + Cr + ΣXi, GEJ 1500	–		4	556	–		–		–		–		–		–		–	
Ni + Cr + ΣXi, GEJ 1610	–		4	556	–		–		–		–		–		–		–	
Ni + Cr + ΣXi, German chromium	1	1018	–		–		–		–		–		–		–		–	
Ni + Cr + ΣXi, Hastelloy R-235	1	1019	4	553	–		–		–		–		–		–		12	1219
Ni + Cr + ΣXi, Hastelloy X	–		–		7	1364	7	1378	–		–		–		–		12	1216 1219
Ni + Cr + ΣXi, Haynes alloy X	–		–		7	1344	–		–		–		–		–		–	
Ni + Cr + ΣXi, Inco 713C	1	1022	4	550	–		–		–		–		–		–		12	1219
Ni + Cr + ΣXi, Inconel	1	1018 1019 1021	–		7	1344 1345 1351 1352 1363 1366	7	1382	7	1396	–		10	299	–		12	1219
Ni + Cr + ΣXi, Inconel, oxidized	–		4	553	9	1314	–		–		–		–		–		–	
Ni + Cr + ΣXi, Inconel 600	1	1018 1019 1021	4	553	7	1344 1345 1351 1352 1363 1366	7	1382	7	1396	–		–		–		12	1216 1220
Ni + Cr + ΣXi, Inconel 625	–		–		–		–		–		–		–		–		12	1220
Ni + Cr + ΣXi, Inconel 702	1	1022	4	553	7	1345 1364	7	1378	–		–		–		–		–	
Ni + Cr + ΣXi, Inconel 713	1	1022	–		–		–		–		–		–		–		–	
Ni + Cr + ΣXi, Inconel 718	–		–		–		–		–		–		–		–		12	1220 1221
Ni + Cr + ΣXi, Inconel 750	–		–		–		–		–		–		–		–		12	1221
Ni + Cr + ΣXi, Inconel B	–		–		7	1350	7	1393	7	1401	–		–		–		–	
Ni + Cr + ΣXi, Inconel Hi C	–		–		–		–		–		–		–		–		12	1220
Ni + Cr + ΣXi, Inconel low C	–		–		–		–		–		–		–		–		12	1220
Ni + Cr + ΣXi, Inconel med C	–		–		–		–		–		–		–		–		12	1220

Substance Name	Thermal Conductivity		Specif. Heat		Thermal Radiative Properties								Thermal Diffusivity		Viscosity		Thermal Expansion	
					Emissivity		Reflectivity		Absorptivity		Transmissiv.							
	V.	Page	V.	Page	V.	Page	V.	Page	V.	Page	V.	Page	V.	Page	V.	Page	V.	Page
Nickel alloys: (continued)																		
Ni + Cr + ΣXi, Inconel X	1	1018	4	553	7	1344 1345 1351 1352 1353 1358 1365	7	1377 1379 139	–		–		10	299	–		12	1221
Ni + Cr + ΣXi, Inconel X750	1	1018	4	553	7	1344 1345 1351 1352 1353 1358 1365	7	1377 1379 139	–		–		–		–		12	1216 1221
Ni + Cr + ΣXi, M-252	1	1022	4	556	7	1366	7	1382	–		–		–		–		12	1221
Ni + Cr + ΣXi, Nichrome	1	1018 1019 1021 1036	–		7	1344 1353	–		–		–		–		–		–	
Ni + Cr + ΣXi, Nichrome V	1	698	4	556	–		–		–		–		–		–		–	
Ni + Cr + ΣXi, Nimocast 713 C	1	1022	–		–		–		–		–		–		–		–	
Ni + Cr + ΣXi, Nimonic 75	1	1019	–		7	1345 1350	–		–		–		–		–		–	
Ni + Cr + ΣXi, Nimonic 75, French	1	1019	–		–		–		–		–		–		–		–	
Ni + Cr + ΣXi, Nimonic 80	1	1018	–		–		–		–		–		–		–		12	1216 1221
Ni + Cr + ΣXi, nimonic 80/80A, French	1	1019	–		–		–		–		–		–		–		–	
Ni + Cr + ΣXi, Nimonic 90	1	1019	–		–		–		–		–		–		–		12	1216 1221
Ni + Cr + ΣXi, Nimonic 95	1	1019	–		–		–		–		–		–		–		–	
Ni + Cr + ΣXi, oxidized	–		–		9	1314	–		–		–		–		–		–	
Ni + Cr + ΣXi, Russian, OKh 20N 60B	–		4	559	–		–		–		–		–		–		–	
Ni + Cr + ΣXi, Russian, OKh 21N 78T	–		4	556	–		–		–		–		–		–		–	
Ni + Cr + ΣXi, Russian, Kh 80T	1	1019	–		–		–		–		–		–		–		–	
Ni + Cr + ΣXi, Russian, EI-435	1	1022	4	559	–		–		–		–		–		–		–	
Ni + Cr + ΣXi, Russian, EI-607	1	1019 1020 1021	–		–		–		–		–		–		–		–	
Ni + Cr + ΣXi, Udimet 500	–		–		7	1365 1412	7	1381 1418	–		–		–		–		–	
Ni + Cr + ΣXi, Udimet 630	–		–		–		–		–		–		–		–		12	1222
Ni + Co + ΣXi	1	1028	–		7	1404 1406 1410	7	1416	–		–		10	302	–		12	1227
Ni + Co + ΣXi, Haynes Stellite 27	1	1029	–		–		–		–		–		–		–		–	
Ni + Co + ΣXi, Konel	–		–		7	1404 1407	–		–		–		–		–		–	
Ni + Co + ΣXi, Nickel 200	1	239 241 1029 1039	–		–		–		–		–		10	122	–		–	
Ni + Co + ΣXi, Nimonic 100	1	1029	–		–		–		–		–		–		–		–	
Ni + Co + ΣXi, Nimonic 105	1	1029	–		–		–		–		–		–		–		–	
Ni + Co + ΣXi, Nimonic 115	1	1029	–		–		–		–		–		–		–		–	

Substance Name	Thermal Conductivity		Specif. Heat		Thermal Radiative Properties								Thermal Diffusivity		Viscosity		Thermal Expansion	
					Emissivity		Reflectivity		Absorptivity		Transmissiv.							
	V.	Page	V.	Page	V.	Page	V.	Page	V.	Page	V.	Page	V.	Page	V.	Page	V.	Page
Nickel alloys: (continued)																		
Ni + Co + ΣXi, Refralloy 26	1	1029	–		–		–		–		–				–		–	
Ni + Co + ΣXi, René 41	1	1022	4	556	7	1353 1365 1366	7	1380 1382	–		–				–		12	1216 1222
Ni + Co + ΣXi, Udiment 700	–		–		–		–		–		–				–		12	1229
Ni + Cu + ΣXi	1	1031	4	562	7	1423 1426	7	1430 1433	7	1436	–		10	303	–		12	1230
Ni + Cu + ΣXi, Corronil	1	1032	–		–		–		–		–				–		–	
Ni + Cu + ΣXi, Monel	1	1032	4	562	7	1423 1428	7	1431	–		–				–		12	1230 1232
Ni + Cu + ΣXi, Monel, 400	1	1032	4	562	7	1423 1428	7	1431	–		–				–		12	1232
Ni + Cu + ΣXi, Monel, 505	1	1032	–		–		–		–		–				–		–	
Ni + Cu + ΣXi, Monel, 506	1	1032	–		–		–		–		–				–		–	
Ni + Cu + ΣXi, Monel, cast	1	1032	–		–		–		–		–				–		–	
Ni + Cu + ΣXi, Monel, H	1	1032	–		–		–		–		–				–		–	
Ni + Cu + ΣXi, Monel, K	1	1032	4	562	7	1428	7	1434	7	1437	–		10	304	–		12	1232
Ni + Cu + ΣXi, Monel, K-500	1	1032	4	562	7	1428	7	1434	7	1437	–		–		–		–	
Ni + Cu + ΣXi, Monel, R	1	1032	–		–		–		–		–				–		–	
Ni + Cu + ΣXi, Monel, R-405	1	1032	–		–		–		–		–				–		–	
Ni + Cu + ΣXi, Monel, S	1	1032	–		–		–		–		–				–		12	1232 1233
Ni + Cu + ΣXi, nickel bronze	1	1032	–		–		–		–		–				–		–	
Ni + Cu + ΣXi, Silicon monel	1	1032	–		–		–		–		–				–		–	
Ni + Fe + ΣXi	1	1035	4	565	7	1439	–		–		–		10	305	–		12	1236
Ni + Fe + ΣXi, contracid	1	1036	–		–		–		–		–				–		–	
Ni + Fe + ΣXi, contracid B 7M	1	1036	–		–		–		–		–				–		–	
Ni + Fe + ΣXi, Hastelloy A	1	1036	–		–		–		–		–				–		–	
Ni + Fe + ΣXi, HyMu-80	1	1036	–		–		–		–		–				–		–	
Ni + Fe + ΣXi, Incoloy 901	–		4	565	–		–		–		–				–		–	
Ni + Fe + ΣXi, Nilo 50	–		–		–		–		–		–				–		12	1238
Ni + Mn + ΣXi	1	1038	4	568	–		–		–		–		10	309	–		12	1241
Ni + Mn + ΣXi, GRD	–		–		–		–		–		–		10	310	–		–	
Ni + Mn + ΣXi, Nickel 211	1	1039	–		–		–		–		–				–		–	
Ni + Mn + ΣXi, Nickel A	1	239 241 1029 1039	–		7	418 419 420 426 427 428 429	7	455	7	471	–		..		–		12	227 1243
Ni + Mn + ΣXi, Nickel D	1	1039	–		–		–		–		–				–		–	
Ni + Mo + ΣXi	1	1041	4	571	7	1442 1446 1450	7	1454	7	1457	–				–		12	1245
Ni + Mo + ΣXi, Hastelloy B	1	1042	4	571	7	1448 1452	7	1455	7	1458	–				.		12	1245 1247
Ni + Mo + ΣXi, Hastelloy C	1	1018	4	556	7	1448 1452	7	1455	7	1458	–		–		–		12	1245 1247

| Substance Name | Thermal Conductivity | | Specif. Heat | | Thermal Radiative Properties | | | | | | | | Thermal Diffusivity | | Viscosity | | Thermal Expansion | |
| | | | | | Emissivity | | Reflectivity | | Absorptivity | | Transmissiv. | | | | | | | |
	V.	Page	V.	Page	V.	Page	V.	Page	V.	Page	V.	Page	V.	Page	V.	Page	V.	Page
Nickel alloys: (continued)																		
Ni + Mo + ΣXi, Hastelloy N		–		–		–		–		–		–		–		–	12	1245 1247
Ni + Mo + ΣXi, Haynes alloy C		–		–	7	1444		–		–		–		–		–		–
Ni + Mo + ΣXi, Inor-b	1	1042		–	7	1444		–		–		–		–		–		–
Ni + Si + ΣXi		–		–		–		–		–		–		–		–	12	1249
Ni + Si + ΣXi, grade A		–		–		–		–		–		–		–		–	12	1249
Ni + Ti + ΣXi		–		–		–		–		–		–	10	307		–		–
Ni + Ti + ΣXi, Permanickel		–		–		–		–		–		–	10	308		–		–
Ni + Al + Mn + ΣXi, alumel	1	1015 1039	4	568		–		–		–		–		–		–		–
Nickel-aluminum intermetallic compounds:																		
NiAl		–		–	8	1316 1318	8	1321		–		–		–		–		–
Ni_3Al		–		–	8	1316 1318	8	1321		–		–		–		–		–
Nickel aluminum oxide, $NiO \cdot Al_2O_3$		–		–		–		–		–		–		–		–	13	484
Nickel antimonide, NiSb	1	1327		–		–		–		–		–		–		–		–
Nickel-antimony intermetallic compound, NiSb	1	1327		–		–		–		–		–		–		–		–
Nickel carbonate, $NiCO_3$		–		–		–		–		–		–		–		–	13	647
Nickel chloride, $NiCl_2$		–	5	863		–		–		–		–		–		–		–
Nickel chloride hexahydrate, $NiCl_2 \cdot 6H_2O$		–	5	866		–		–		–		–		–		–		–
Nickel fluoride, NiF_2		–	5	973		–		–		–		–		–		–	13	1052
Nickel fluosilicate hexahydrate (A)		–	5	966		–		–		–		–		–		–		–
Nickel fluosilicate hexahydrate (B)		–	5	970		–		–		–		–		–		–		–
Nickel iron oxide, $NiO \cdot Fe_2O_3$		–	5	1530		–		–		–		–		–		–		–
Nickel iron oxide, nonstoichiometric		–	5	1533		–		–		–		–		–		–		–
Nickel L	1	238 239		–		–		–		–		–		–		–	12	227
Nickel-niobium intermetallic compound, Ni_3Nb		–		–		–		–		–		–		–		–	12	596
Nickel O	1	239		–		–		–		–		–		–		–		–
$NiO + Al_2O_3$ powders		–		–	8	556		–		–		–		–		–		–
Nickel oxide + nickel aluminum compound + ΣXi, cermet		–		–	8	1393 1394	8	1398		–		–		–		–		–
Nickel oxides:																		
NiO	2	171	5	172	8	337 339	8	342		–		–		–		–	13	319
Ni_2O_3		–		–		–	8	342		–		–		–		–		–
Nickel S		–		–		–		–		–		–		–		–	12	228
Nickel selenide, $NiSe_2$		–	5	549		–		–		–		–		–		–		–
Nickel selenide, nonstoichiometric		–	5	545		–		–		–		–		–		–		–
Nickel silicate, Ni_2SiO_4		–		–		–		–		–		–		–		–	13	727

124

Substance Name	Thermal Conductivity		Specif. Heat		Thermal Radiative Properties								Thermal Diffusivity		Viscosity		Thermal Expansion	
					Emissivity		Reflectivity		Absorptivity		Transmissiv.							
	V.	Page	V.	Page	V.	Page	V.	Page	V.	Page	V.	Page	V.	Page	V.	Page	V.	Page
Nickel silicides:																		
Ni₂Si		–		–		–		–		–		–		–		–	13	1212
NiSi₂		–		–	8	1173		–		–		–		–		–		–
Nickel sulfate hexahydrate		–	5	1206		–												–
Nickel sulfides:																		
NiS		–	5	693		–		–		–		–		–		–		–
Ni₃S₂	2	705	5	696		–		–		–		–		–		–		–
Nickel telluride, NiTe₂		–	5	738		–		–		–		–		–		–		–
Nickel telluride, nonstoichiometric		–	5	735		–		–		–		–		–		–		–
Nickel + titanium boride, cermet		–		–		–	8	1435		–		–		–		–		–
Nickel-titanium intermetallic compound, NiTi		–		–		–		–		–		–		–		–	12	597
Nickel + titanium nickel compound, mixture	1	1436		–		–		–		–		–		–		–		–
Nickel titanium oxide, NiO·TiO₂		–		–		–		–		–		–		–		–	13	569
Nickel-yttrium intermetallic compound, Ni₅Y		–		–		–		–		–		–		–		–	12	598
Nickel zinc ferrate, NiZnFe₂O₄	2	298		–		–		–		–		–		–		–		–
Nickel zinc ferrite		–	5	1536		–		–		–		–		–		–		–
Nickel zinc iron oxide, nonstoichiometric		–	5	1536		–		–		–		–		–		–		–
Nickelous oxide		–	5	172		–		–		–		–		–		–		–
Nil Alba, ZnO	2	243		–		–		–		–		–		–		–		–
Niobium, Nb	1	245	4	153	7	474 480 482 486	7	492	7	497		–	10	125		–	12	236
Niobium alloys:																		
Nb + Mo		–		–		–		–		–		–		–		–	12	900
Nb + Re		–		–		–		–		–		–		–		–	12	949
Nb + Ta		–		–		–		–		–		–	10	251		–		–
Nb + U	1	713		–		–		–		–		–		–		–	12	954
Nb + V		–		–		–		–		–		–		–		–	12	955
Nb + W		–		–	7	981 984		–		–		–		–		–		–
Nb + Zr	1	716	4	422	7	988 992 994		–		–		–		–		–	12	958
Nb + Fe + ΣXi		–	4	574		–		–		–		–		–		–		–
Nb + Fe + ΣXi, Russian, ferroniobium		–	4	574		–		–		–		–		–		–		–
Nb + Mo + ΣXi	1	1046	4	577	7	1460		–		–		–	10	312		–	12	1250
Nb + Ta + ΣXi	1	1049	4	580	7	1463 1466		–		–		–	10	314		–	12	1253
Nb + Ti + ΣXi	1	1052	4	583		–		–		–		–	10	316		–	12	1257
Nb + W + ΣXi	1	1055	4	586	7	1469		–		–		–	10	318		–	12	1259
Niobium-aluminum intermetallic compound, NbAl₃		–		–		–	8	1322		–		–		–		–		–
Niobium beryllide, NbBe₁₂		–	5	319		–		–		–		–		–		–		–
Niobium boride, NbB₂		–	5	365	8	697 699 701		–		–		–		–		–	13	767

| Substance Name | Thermal Conductivity | | Specif. Heat | | Thermal Radiative Properties | | | | | | | | | | Thermal Diffusivity | | Viscosity | | Thermal Expansion | |
| | | | | | Emissivity | | Reflectivity | | Absorptivity | | Transmissiv. | | | | | | | | | |
	V.	Page	V.	Page	V.	Page	V.	Page	V.	Page	V.	Page	V.	Page	V.	Page	V.	Page
Niobium boride, nonstoichiometric		–	5	361		–		–		–		–		–		–		–
Niobium carbides:																		
NbC	2	582	5	442	8	785 787 789		–		–		–	10	474		–	13	858
Nb_2C		–		–		–		–		–		–		–		–	13	936
Nonstoichiometric		–	5	439		–		–		–		–		–		–		–
Niobium fluoride, NbF_5		–	5	976		–		–		–		–		–		–		–
Niobium nitrides:																		
NbN		–		–	8	1087		–		–		–		–		–	13	1162
Nb_2N		–		–	8	1087		–		–		–		–		–		–
Niobium oxides:																		
NbO		–	5	175		–		–		–		–		–		–		–
NbO_2		–	5	178		–		–		–		–		–		–		–
Nb_2O_5		–	5	181	8	347	8	349		–		–		–		–	13	323
Niobium phosphate, $NbPO_5$		–		–		–		–		–		–		–		–	13	690
Niobium silicides:																		
$NbSi_2$				–	8	1173		–		–		–		–		–	13	1213
Nb_5Si_3		–		–		–		–		–		–		–		–	13	1213
Niobium tantalum oxide, $2Nb_2O_5 \cdot Ta_2O_5$		–		–		–		–		–		–		–		–	13	543
Niobium-tin intermetallic compound, Nb_3Sn		–		–		–		–		–		–		–		–	12	601
Niobium vanadium oxide, $Nb_2O_5 \cdot V_2O_5$		–		–		–		–		–		–		–		–	13	596
Niton	3	84		–		–		–		–		–		–		–		–
Nitric oxide, NO	3	106	6	83		–		–		–		–		–	11	82		–
Nitric oxide-nitrogen, mixture		–		–		–		–		–		–		–	11	495		–
Nitric oxide-nitrous oxide, mixture		–		–		–		–		–		–		–	11	492		–
m-Nitroaniline		–	6s	69		–		–		–		–		–		–		–
o-Nitroaniline		–	6s	69		–		–		–		–		–		–		–
p-Nitroaniline		–	6s	69		–		–		–		–		–		–		–
m-Nitrobenzoic acid		–	6s	69		–		–		–		–		–		–		–
o-Nitrobenzoic acid		–	6s	69		–		–		–		–		–		–		–
p-Nitrobenzoic acid		–	6s	69		–		–		–		–		–		–		–
Nitrobenzol		–	6s	69		–		–		–		–		–		–		–
Nitrocarbol		–	6s	69		–		–		–		–		–		–		–
m-Nitrodracylic acid		–	6s	69		–		–		–		–		–		–		–
o-Nitrodracylic acid		–	6s	69		–		–		–		–		–		–		–
p-Nitrodracylic acid		–	6s	69		–		–		–		–		–		–		–
Nitrogen, N_2	3	64	6	39		–		–		–		–	10	128	11	48		–
Nitrogen, monatomic		–	6s	69		–		–		–		–		–		–		–
Nitrogen oxides:																		
NO_2	3	108	6	90		–		–		–		–		–	11	85		–
N_2O	3	114		–		–		–		–		–		–		–		–
Nitrogen-oxygen, mixture	3	434		–		–		–		–		–		–	11	497		–

Substance Name	Thermal Conduc- tivity		Specif. Heat		Thermal Radiative Properties								Thermal Diffu- sivity		Visco- sity		Thermal Expan- sion	
					Emis- sivity		Reflec- tivity		Absorp- tivity		Trans- missiv.							
	V.	Page	V.	Page	V.	Page	V.	Page	V.	Page	V.	Page	V.	Page	V.	Page	V.	Page
Nitrogen-propane, mixture	3	438	–		–		–		–		–		–		–		–	
Nitrogen-steam, mixture	3	468	–		–		–		–		–		–		–		–	
Nitrogen-xenon, mixture	3	377	–		–		–		–		–		–		–		–	
Nitromethane	–		6s	69	–		–		–		–		–		–		–	
Nitrophenol	2	1001	–		–		–		–		–		–		–		–	
Nitroso elastomer compound	–		–		–		–		–		–		–		–		13	1520
Nirtous oxide, N_2O	3	114	6	92	–		–		–		–		–		11	87	–	
Nitrous oxide-propane, mixture	–		–		–		–		–		–		–		11	499	–	
Nitrous oxide-sulfur dioxide, mixture	–		–		–		–		–		–		–		11	536	–	
Nivac	1	238	–		–		–		–		–		–		–		–	
n-Nonane	3	230	6	261	–		–		–		–		–		–		–	
Nonoxides, mixture	–		–		–		–		–		–		10	517	–		–	
Normal phosphate	–		–		–		–		–		–		–		–		13	687
Nylon	2	945	–		–		–		–		–		10	595	–		–	
Nylon, miscellaneous polymers	–		–		–		–		–		–		–		–		13	1498
Nylon, phenolic, low density polymers	–		–		–		–		–		–		–		–		13	1499
Nylon 6	2	959	–		–		–		–		–		–		–		13	1499
Nylon 610	–		–		–		–		–		–		–		–		13	1499
Oak, white	2	1082	–		–		–		–		–		–		–		–	
Obsidian	–		–		–		8	1681	–		–		–		–		–	
Octafluorocyclobutane	–		–		–		–		–		–		–		11	199	–	
n-Octane	3	233	6	266	–		–		–		–		–		11	204	–	
2-Octyl acetate	–		6s	70	–		–		–		–		–		–		–	
n-Octyl alcohol	–		6s	70	–		–		–		–		10	590	–		–	
2-Octyl ethanoate	–		6s	70	–		–		–		–		–		–		–	
Oil, castor	–		–		–		–		–		–		10	632	–		–	
Oil, Mirbane	1	1022	6s	69	–		–		–		–		–		–		–	
Olefiant gas	–		6	185	–		–		–		–		–		–		–	
Oligoclase, silicate mineral	–		–		–		8	1689	–		–		–		–		–	
Olivine	2	275	–		–		8	1689	–		–		10	417 427	–		–	
Opal	–		·		–		–		–		8	421	–		–		–	
Opal glass	–		–		–		8	1646	–		–		–		–		–	
Optical, T-12 commercial material	–		–		–		8	1451	–		8	1453	–		–		–	
Orange marmalade	–		–		–		–		–		–		10	642	–		–	
Osmium, Os	1	254	4	157	7	500	–		–		–		10	130	–		12	244
Oxalic acid dinitrile	–		6s	24	–		–		–		–		–		··		–	
Oxalic ester	–		6s	30	–		–		–		–		–		–		–	
Oxalonitrile	–		6s	24	–		–		–		–		–		–		–	
Oxides, mixture	–		–		–		–		–		–		10	425	–		–	
Oxirane	–		6s	37	–		–		–		–		–		–		–	
Oxomethane	–		6s	42	–		–		–		–		–		–		–	
1,1'-Oxydi-2-propanediol	–		6s	34	–		–		–		–		–		–		–	

Substance Name	Thermal Conductivity		Specif. Heat		Thermal Radiative Properties								Thermal Diffusivity		Viscosity		Thermal Expansion	
					Emissivity		Reflectivity		Absorptivity		Transmissiv.							
	V.	Page	V.	Page	V.	Page	V.	Page	V.	Page	V.	Page	V.	Page	V.	Page	V.	Page
Oxyethylene		–	6s	37		–		–		–		–		–		–		–
Oxygen, O$_2$	3	76	6	48		–		–		–		–		–	11	56		–
Oxygen, monatomic		–	6s	70		–		–		–		–		–		–		–
Oxygen + xenon, mixture	3	379		–		–		–		–		–		–		–		–
Oxygen fluoride		–	6s	71		–		–		–		–		–		–		–
Oxymethylene		–	6s	42		–		–		–		–		–		–		–
Oxyphenic acid		–	6s	83		–		–		–		–		–		–		–
Packed beds, nonmetallic		–		–		–		–		–		–	10	562		–		–
Paint, AISI 99 grey		–		–		–		–	9	485	9	486		–		–		–
Paint, black Boysen no. 11		–		–		–	9	487		–		–		–		–		–
Paint, black velvet, 3M		–		–	9	533 535 536	9	538	9	540		–		–		–		–
Paint, Channel black		–		–		–		–		–		–	10	22 35		–		–
Paint, Cat-A-Lac white		–		–	9	491		–	9	492 493		–		–		–		–
Paint, Cat-A-Lac black		–		–	9	82 84	9	86	9	89		–		–		–		–
Paint, CM-145 Al$_2$O$_3$		–		–		–	9	33	9	42		–		–		–		–
Paint, CM-146 TiO$_2$		–		–		–		–	9	274		–		–		–		–
Paint, CM-147 ZnO		–		–		–		–	9	378		–		–		–		–
Paint, Dulite 1015		–		–	9	496		–		–		–		–		–		–
Paint, Dulite II		–		–	9	496		–		–		–		–		–		–
Paint, Dutch Boy 46H 47		–		–	9	499	9	500		–		–		–		–		–
Paint, Eastman white reflecting		–		–		–	9	62		–		–		–		–		–
Paints, Fuller:																		
D-70-6342		–		–	9	502		–		–		–		–		–		–
Flat black decoret		–		–		–	9	504		–		–		–		–		–
Flat black silicone		–		–	9	81		–	9	89		–		–		–		–
Harvard gray No. 2946		–		–		–	9	504		–		–		–		–		–
Light brown No. 2909		–		–		–	9	504		–		–		–		–		–
Mariposa blue decoret No. 2889		–		–		–	9	504		–		–		–		–		–
TL-8606 No. 43		–		–		–	9	504		–		–		–		–		–
TL-9465 No. 45		–		–		–	9	504		–		–		–		–		–
Velvet black No. 1518		–		–		–	9	504		–		–		–		–		–
No. 171-A-152		–		–	9	3		–	9	21		–		–		–		–
No. 171-W-560		–		–	9	501		–	9	507		–		–		–		–
No. 172-A-1		–		–		–	9	13		–		–		–		–		–
No. 517-W-1		–		–	9	212		–	9	256 263		–		–		–		–
No. 517-W-7		–		–	9	211	9	227		–		–		–		–		–
Paint, Glasurit white epoxy		–		–		–		–	9	508		–		–		–		–

Substance Name	Thermal Conductivity		Specif. Heat		Thermal Radiative Properties								Thermal Diffusivity		Viscosity		Thermal Expansion	
					Emissivity		Reflectivity		Absorptivity		Transmissiv.							
	V.	Page	V.	Page	V.	Page	V.	Page	V.	Page	V.	Page	V.	Page	V.	Page	V.	Page
Paints, Glidden:	-		-		-		9	510	-		-		-		-		-	
Black lacquer No. 131-B-190B	-		-		-		9	510	-		-		-		-		-	
Black lacquer No. 9099	-		-		-		9	510	-		-		-		-		-	
Flat black lacquer No. 131-B-216	-		-		-		9	510	-		-		-		-		-	
Flat white No. 2995	-		-		-		9	510	-		-		-		-		-	
Flat white No. 5064	-		-		-		9	510	-		-		-		-		-	
Japalac black enamel No. 1207	-		-		-		9	510	-		-		-		-		-	
Japalac flat black No. 1208	-		-		-		9	510	-		-		-		-		-	
Missile black No. RGL-22818	-		-		-		9	510	-		-		-		-		-	
Paint, glidden Zapon black	-		-		-		9	510 572	-		-		-		-		-	
Paint, Goddard 78-2B white	-		-		-		-		9	273	-		-		-		-	
Paints, glyptal:																		
Black	-		-		9	513	9	515	-		-		-		-		-	
Brown	-		-		-		9	515	-		-		-		-		-	
Red	-		-		-		9	510	-		-		-		-		-	
Paint, Kry-Kote white	-		-		9	523	-		-		-		-		-		-	
Paints, krylon:																		
Black	-		-		9	524	9	526	-		-		-		-		-	
Silver aluminum	-		-		-		9	526	-		-		-		-		-	
White	-		-		-		9	526	-		-		-		-		-	
Yellow	-		-		-		9	526	-		-		-		-		-	
Paint, LMSC/Dow Corning Thermatrol	-		-		9	211	-				9	257 270	-		-		-	
Paint, No. 1019 GAFC black epoxy	-		-		9	517	-		-		-		-		-		-	
Paint, North American aviation No. 211.2.1	-		-		9	82	9	86	-		-		-		-		-	
Paint, Nuclear enterprise No. 561	-		-		-		9	543	-		-		-		-		-	
Paint, Parsons black	-		-		9	545 546	9	548	9	550	-		-		-		-	
Paint, Pedigree red	-		-		-		9	551	-		-		-		-		-	
Paint, proven flate white SP-15	-		-		-		9	553	-		-		-		-		-	
Paint, Pyromark OAO	-		-		-		9	244	-		-		-		-		-	
Paint, Pyromark standard black	-		-		9	117 119 123	-		-		-		-		-		-	
Paint, Pyromark TiO$_2$	-		-		-		9	224	9	272	-		-		-		-	
Paints, Rust-Oleium:																		
No. 960	-		-		-		9	300	-		-		-		-		-	
Fire hydrant red no. 1210	-		-		-		9	555	-		-		-		-		-	
Green no. 205	-		-		-		9	555	-		-		-		-		-	
Red no. 215	-		-		-		9	555	-		-		-		-		-	
Silver gray no. 208	-		-		-		9	555	-		-		-		-		-	
White no. 225	-		-		-		9	555	-		-		-		-		-	
Paint, satin walnut	2	1089	-		-		-		-		-		-		-		-	

Substance Name	Thermal Conductivity		Specif. Heat		Thermal Radiative Properties								Thermal Diffusivity		Viscosity		Thermal Expansion	
					Emissivity		Reflectivity		Absorptivity		Transmissiv.							
	V.	Page	V.	Page	V.	Page	V.	Page	V.	Page	V.	Page	V.	Page	V.	Page	V.	Page
Paints, Sherwin-Williams:																		
Enameloid flat black		–		–		–	9	559		–		–		–		–		–
Moroon enamel		–		–	9	557		–		–		–		–		–		–
Kemtone Shasta white no. 793		–		–		–	9	559		–		–		–		–		–
Silverbrite no. 55		–		–	9	6		–		–		–		–		–		–
Paint, sicon black		–		–	9	561	9	563		–		–		–		–		–
Paint, Spraint grey no. 63		–		–		–	9	564		–		–		–		–		–
Paints, Sprayon:																		
High-heat no. 324		–		–		–	9	566		–		–		–		–		–
Machinery dark grey enamel no. 325		–		–		–	9	566		–		–		–		–		–
Machinery light grey enamel no. 326		–		–		–	9	566		–		–		–		–		–
Paint, Sylvania phosphor		–		–	9	568		–		–		–		–		–		–
Paint, thermal black		–		–		–		–		–		–	10	22 35		–		–
Paint, trial aluminum No. 18270		–		–	9	6		–		–		–		–		–		–
Paint, U. S. Army olive drab		–		–		–	9	569		–		–		–		–		–
Paint, vita-var grey		–		–		–		–	9	570	9	571		–		–		–
Paint, white velvet, 3M		–		–		–	9	538		–		–		–		–		–
Paint, zinc oxide pigmented Z-93		–		–	9	302 304	9	314	9	360 372 392		–		–		–		–
Palladium, Pd	1	258	4	160	7	502 504 507 510	7	512	7	515 518	7	520	10	131		–	12	248
Palladium alloys:																		
Pd + Ag	1	727	4	425		–		–		–		–		–		–	12	962
Pd + Au	1	723		–		–		–		–		–		–		–	12	812
Pd + Co		–		–		–		–		–		–		–		–	12	747
Pd + Cu	1	720		–		–		–		–		–		–		–		–
Pd + Ni		–		–		–		–		–		–		–		–	12	926
Pd + Pt	1	726		–		–		–		–		–		–		–		–
Pd + Au + ΣXi		–		–		–		–		–		–		–		–	12	1265
Palladium-indium intermetallic compound, PdIn		–		–		–	8	1352		–		–		–		–		–
Palladium tellurides:																		
PdTe		–	5	741		–		–		–		–		–		–		–
PdTe$_2$		–	5	744		–		–		–		–		–		–		–
Palladium-tin intermetallic compound, Pd$_3$Sn		–		–		–		–		–		–		–		–	12	605 607 608
Palladium-ytterbium intermetallic compound, Pd$_3$Yb		–		–		–		–		–		–		–		–	12	606 607 609
Pallagold, gold alloy		–		–		–		–		–		–		–		–	12	814
Paper	2	1127		–		–		–		–		–		–		–		–
Paraplex P-444A		–		–		–		–		–		–		–		–	13	1421

Substance Name	Thermal Conductivity		Specif. Heat		Thermal Radiative Properties								Thermal Diffusivity		Viscosity		Thermal Expansion	
					Emissivity		Reflectivity		Absorptivity		Transmissiv.							
	V.	Page	V.	Page	V.	Page	V.	Page	V.	Page	V.	Page	V.	Page	V.	Page	V.	Page
Pentadecane		–	6s	72		–		–		–		–		–		–		–
n-Pentane	3	236	6	272		–		–		–		–		–	11	206		–
tert-Pentanol		–	6s	61		–		–		–		–		–		–		–
1-Pentanol		–	6s	72		–		–		–		–		–		–		–
3-Pentanol		–	6s	72		–		–		–		–		–		–		–
3-Pentanone		–	6s	72		–		–		–		–		–		–		–
3-Pentanone, metacetone		–	6s	72		–		–		–		–		–		–		–
1-Pentene		–	6s	73		–		–		–		–		–		–		–
2-Pentene		–	6s	73		–		–		–		–		–		–		–
1-Pentine		–	6s	73		–		–		–		–		–		–		–
2-Pentine		–	6s	73		–		–		–		–		–		–		–
Pentylene		–	6s	73		–		–		–		–		–		–		–
1-Pentyne		–	6s	73		–		–		–		–		–		–		–
2-Pentyne		–	6s	73		–		–		–		–		–		–		–
Perchloroethylene		–	6s	90		–		–		–		–		–		–		–
Perchloromethane		–	6	159		–		–		–		–		–		–		–
Perchlorovinyl		–	6s	90		–		–		–		–		–		–		–
Perfluoroethane		–	6s	44		–		–		–		–		–		–		–
Periclase	2	160		–		–		–		–		–		–		–		–
Perlite	2	827		–		–		–		–		–		–		–		–
Perovskite fluorides		–		–		–	8	976 980 984 988		–	8	978 982 986 990		–		–		–
Peroxide		–	6s	49		–		–		–		–		–		–		–
Petalite	2	935		–		–		–		–		–		–		–		–
Phenanthren	2	1004		–		–		–		–		–		–		–		–
Phene, hexadeuterated		–	6s	2		–		–		–		–		–		–		–
Phenol-formaldehyde plastic, reinforced		–		–		–		–		–		–	10	575		–		–
Phenoxybenzene		–	6s	73		–		–		–		–		–		–		–
Phenyl amine		–	6s	1		–		–		–		–		–		–		–
Phenyl bromide		–	6s	4		–		–		–		–		–		–		–
1-Phenyl butane		–	6s	11		–		–		–		–		–		–		–
Phenyl carbinol		–	6s	2		–		–		–		–		–		–		–
Phenyl chloride		–	6s	21		–		–		–		–		–		–		·
Phenyl ethane		–	6s	35		··		–		–		–		–		–		–
Phenyl ether	2	990	6s	73		–		–		–		–		–		–		–
Phenyl ethylene		–	6s	84		–		–		–		–		–		–		–
Phenyl fluoride		–	6s	41		–		–		–		–		–		–		–
Phenyl formic acid		–	6s	2		–		–		–		–		–		–		–
Phenyl iodide		–	6s	55		–		–		–		–		–		–		–
Phenyl methane		–	6	285		–		–		–		–		–		–		–
1-Phenyl propane		–	6s	80		–		–		–		–		–		–		–
2-Phenyl propane		–	6s	22		–		–		–		–		–		–		–

Substance Name	Thermal Conduc- tivity		Specif. Heat		Thermal Radiative Properties										Thermal Diffu- sivity		Visco- sity		Thermal Expan- sion	
					Emis- sivity		Reflec- tivity		Absorp- tivity		Trans- missiv.									
	V.	Page	V.	Page	V.	Page	V.	Page	V.	Page	V.	Page	V.	Page	V.	Page	V.	Page		
Phlogopite		–		–		–		–		–		–	10	548		–		–		
Phosgene		–	6s	74		–		–		–		–		–		–		–		
Phosphates, miscellaneous		–		–		–	8	607		–		–		–		–		–		
Phosphides, miscellaneous		–		–	8	1104	8	1106		–		–		–		–		–		
Phosphine		–	6s	74		–		–		–		–		–		–		–		
Phosphine, trideuterated		–	6s	75		–		–		–		–		–		–		..		
Phosphoretted hydrogen		–	6s	74		–		–		–		–		–		–		–		
Phosphorus	2	86	5	18		–		–		–		–		–		–		–		
Phosphorus, black		–	5	18		–		–		–		–		–		–		–		
Phosphorus, white		–		–		–		–		–		–	10	134		–		–		
Phosphorus chloride, PCl₃		–	5	869		–		–		–		–		–		–		–		
			6s	75																
Phosphorus fluoride, PF₃		–	6s	75		–		–		–		–		–		–		–		
Phosphorus sulfide, P₄S₃		–		–		–		–		–		–		–		–	13	1240		
Phosphorusous chloride		–	5	869		–		–		–		–		–		–		–		
Phosphuretted hydrogen		–	6s	74		–		–		–		–		–		–		–		
Picrochromite		–		–		–		–		–		–		–		–	13	486		
Pines, pitch	2	1083		–		–		–		–		–		–		–		–		
Pines, white	2	1083		–		–		–		–		–		–		–		–		
Pladuram	1	416		–		–		–		–		–		–		–		–		
Plastics, vinyl		–		–		–	8	1741	8	1743	8	1745		–		–		–		
Plaster	2	887		–		–		–		–		–		–		–		–		
Plastic		–		–		–		–		–		–	10	553		–		–		
														555						
														556						
Plastic, silicone-asbestos		–		–		–		–		–		–	10	558		–		–		
Platinoid	1	981		–		–		–		–		–		–		–		–		
Platinum, Pt	1	262	4	163	7	524	7	544	7	551		–	10	135		–	12	254		
						529		547		554										
						532		549		557										
						536														
Platinum, NBS		–		–	7	538		–	7	552		–		–		–		–		
Platinum alloys:																				
Pt + Ag	1	745		–		–		–		–		–		–		–		–		
Pt + Au	1	733		–		–		–		–		–		–		–	12	817		
Pt + Cu	1	730		–		–		–		–		–		–		–		–		
Pt + Fe		–		–		–		–		–		–		–		–	12	861		
Pt + Ir	1	734		–		–		–		–		–		–		–	12	836		
Pt + Ni		–		–		–		–		–		–		–		–	12	930		
Pt + Pd	1	737		–		–		–		–		–		–		–		–		
Pt + Rh	1	738		–	7	997		–		–		–		–		–	12	967		
						1000														
						1004														
Pt + Rh, Pt-10Rh		–		–	7	997		–		–		–		–		–		–		
Pt + Rh, Pt-13Rh		–		–	7	1002		–		–		–		–		–		–		
						1005														
Pt + Ru	1	743		–		–		–		–		–		–		–	12	970		

132

Substance Name	Thermal Conductivity V.	Page	Specif. Heat V.	Page	Thermal Radiative Properties Emissivity V.	Page	Reflectivity V.	Page	Absorptivity V.	Page	Transmissiv. V.	Page	Thermal Diffusivity V.	Page	Viscosity V.	Page	Thermal Expansion V.	Page
Platinum sulfides:																		
PtS		–	5	699		–		–		–		–		–		–		–
PtS$_2$		–	5	702		–		–		–		–		–		–		–
Platinum telluride:																		
PtTe		–	5	747		–		–		–		–		–		–		–
PtTe$_2$		–	5	750		–		–		–		–		–		–		–
Platinum wire, grade MPTU 4292-53		–		–	7	526		–		–		–		–		–		–
Plexiglas	2	960		–		–		–		–	8	1724	10	603		–		–
Plexiglas AN-P-44A	2	961		–		–		–		–		–		–		–		–
Plioforam	2	950		–		–		–		–		–		–		–		–
Pluton cloth	2	1100		–		–		–		–		–		–		–		–
Plutonium, Pu	1	270	4	167		–		–		–		–	10	142		–	12	260
Plutonium, α	1	271		–		–		–		–		–		–		–		–
Plutonium alloys:																		
Pu + Al	1	746		–		–		–		–		–		–		–	12	660
Pu + Al, delta-stabilized	1	746		–		–		–		–		–		–		–		–
Pu + Fe	1	747		–		–		–		–		–		–		–		–
Pu + Ga		–		–		–		–		–		–		–		–	12	807
Pu + Ce + ΣXi		–	4	589		–		–		–		–		–		–		–
Plutonium carbides:																		
PuC		–	5	445		–		–		–		–	10	476		–	13	866
Pu$_2$C$_3$		–		–		–		–		–		–		–		–	13	936
Plutonium carbide + uranium carbide, mixture		–		–		–		–		–		–	10	532		–		–
Plutonium dioxide, PuO$_2$		–	5	190		–		–		–		–		–		–	13	331
Plutonium dioxide + uranium dioxide, mixture		–		–		–		–		–		–	10	442		–		–
Plutonium nitride, PuN		–		–		–		–		–		–		–		–	13	1162
Plutonium nitride + uranium nitride, mixture		–		–		–		–		–		–		–		–	13	1163
Plutonium phosphide, PuP		–		–		–		–		–		–		–		–	13	1176
Plutonium sulfide, PuS		–		–		–		–		–		–		–		–	13	1227
Poly(acetyl triallyl citrate)		–		–		–		–		–		–		–		–	13	1384
Polyacrylonitrile		–		–		–		–		–		–		–		–	13	1385
Poly(acrylonitrile-butadiene)		–		–		–		–		–		–		–		–	13	1386
Poly(allyl diglycol carbonate)		–		–		–		–		–		–		–		–	13	1388
Poly(allyl methacrylate)		–		–		–		–		–		–		–		–	13	1389
Poly(butadiene-styrene)		–		–		–		–		–		–		–		–	13	1394
Poly(butyl methacrylate)		–		–		–		–		–		–		–		–	13	1396
Polycarbonate		–		–		–		–		–		–		–		–	13	1403
Polycarbonate, makrolon		–		–		–		–		–		–		–		–	13	1405
Polycarbonate, merlon		–		–		–		–		–		–		–		–	13	1405
Polychloroethylene, polyvinyl chloride	2	953		–		–		–		–		–		–		–		–
Polychloroethylene, polyvinyl chloride, plasticized	2	954		–		–		–		–		–		–		–		–

Substance Name	Thermal Conductivity		Specif. Heat		Thermal Radiative Properties								Thermal Diffusivity		Viscosity		Thermal Expansion	
					Emissivity		Reflectivity		Absorptivity		Transmissiv.							
	V.	Page	V.	Page	V.	Page	V.	Page	V.	Page	V.	Page	V.	Page	V.	Page	V.	Page
Poly(chlorotrifluoroethylene)	2	970	–		–		–		–		–			–			13	1409
Poly(chlorotrifluoroethylene-1,1-difluoroethylene)		–		–		–		–		–		–		–		–	13	1413
Poly(chlorostyrene)		–		–		–		–		–		–		–		–	13	1415
Poly(cyclohexyl methacrylate)		–		–		–		–		–		–		–		–	13	1455
Poly(n-decyl methacrylate)		–		–		–		–		–		–		–		–	13	1418
Poly(diallyl phenylphosphonate)		–		–		–		–		–		–		–		–	13	1390
Poly(diallyl phthalate)		–		–		–		–		–		–		–		–	13	1391
Polyester		–		–		–		–		–		–		10	596	–	13	1419
Polyester resin, selectron 5026		–		–		–		–		–		–		–		–	13	1512
Poly(ethyl acrylate)		–		–		–		–		–		–		–		–	13	1423
Poly(ethyl acrylate), pseudo balsa	2	983	–		–		–		–		–		–			–		–	
Poly(ethyl methacrylate)		–		–		–		–		–		–		–		–	13	1424
Polyethylene	2	956	–		–		–		–		8	1705	10	597	–	13	1427
Polyethylene, chlorosulfonated	2	983	–		–		–		–		–		–			–		–	
Polyethylene, Marlex 50		–		–		–		–		–		–		–		–	13	1430
Polyethylene, Marlex 5095		–		–		–		–		–		–		–		–	13	1429
Polyethylene, Marlex 6001		–		–		–		–		–		–		–		–	13	1429
Polyethylene, Marlex 6002		–		–		–		–		–		–		–		–	13	1429
Polyethylene, Marlex 6009		–		–		–		–		–		–		–		–	13	1429
Polyethylene chlorosulfonate		–		–		–		–		–		–		–		–	13	1434
Polyethylene glycol dimethacrylate		–		–		–		–		–		–		–		–	13	1435
Poly(ethylene glycol-terephthatic acid), mylar		–		–		–		–		8	1708 1710	8	1711	–		–		–	
Polyethylene terephthalate		–		–		–		–		–		–		–		–	13	1437
Polyethylene terephthalene		–		–		–		–		–		–		10	600	–		–	
Polyflon M-11		–		–		–		–		–		–		–		–	13	1445
Polyhexahydro-2H-azepine-2-one, silon	2	959	–		–		–		–		–		–			–		–	
Poly(n-hexl methacrylate)		–		–		–		–		–		–		–		–	13	1453
Poly(2-hydroxyethyl methacrylate)		–		–		–		–		–		–		–		–	13	1456
Poly(2-hydroxypropyl methacrylate)		–		–		–		–		–		–		–		–	13	1458
Polyimide, kapton		–		–		–		–		8	1714	8	1716	–		–		–	
Polyimide, kapton film		–		–		–		–		8	1714	8	1716	–		–		–	
Polyisobutylene		–		–		–		–		–		–		–		–	13	1402
Polymer		–		–		–		–		–		–		10	593	–		–	
Polymer, catalin		–		–		–		–		–		–		–		–	13	1512
Polymer, PEMA		–		–		–		–		–		–		–		–	13	1424
Polymer, PHEMA		–		–		–		–		–		–		–		–	13	1456
Polymer, PMA		–		–		–		–		–		–		–		–	13	1466
Polymer, PMEMA		–		–		–		–		–		–		–		–	13	1464
Polymer, PnOMA		–		–		–		–		–		–		–		–	13	1482
Polymer, PnPMA		–		–		–		–		–		–		–		–	13	1484
Polymer, resibond 907		–		–		–		–		–		–		–		–	13	1504

Substance Name	Thermal Conductivity		Specif. Heat		Thermal Radiative Properties								Thermal Diffusivity		Viscosity		Thermal Expansion	
					Emissivity		Reflectivity		Absorptivity		Transmissiv.							
	V.	Page	V.	Page	V.	Page	V.	Page	V.	Page	V.	Page	V.	Page	V.	Page	V.	Page
Polymethacrylate		–		–		–		–		–		–	10	601		–		–
Poly(2-methoxyethyl methacrylate)		–		–		–		–		–		–		–		–	13	1464
Poly(methyl acrylate)		–		–		–		–		–		–		–		–	13	1466
Poly(2-methyl 1,3-butadiene 2-methyl propene)		–		–		–		–		–		–		–		–	13	1468
Poly(methyl methacrylate)	2	960		–		–	8	1719	8	1721	8	1723	10	602		–	13	1470
Poly(methyl methacrylate), AN-P-44A	2	961		–		–		–		–		–		–		–		–
Poly(methyl methacrylate), Perspex	2	961		–		–		–		–		–		–		–		–
Poly(methyl pentene)		–		–		–		–		–		–		–		–	13	1475
Poly(2-methyl 5-vinyl pyridine-acrylonitrile butadiene)		–		–		–		–		–		–		–		–	13	1476
Polymethylene		–		–		–		–		–		–		–		–	13	1477
Polyoxymethylene		–		–		–		–		–		–		–		–	13	1478
Poly(n-octyl methacrylate)		–		–		–		–		–		–		–		–	13	1482
Poly(n-propyl methacrylate)	2	101		–		–		–		–		–		–		–		–
Polypropylene		–		–		–		–		–		–	10	604		–	13	1486
Poly(pyromellitimide)		–		–		–		–		–		–		–		–	13	1462
Polystyrene		–		–		–	8	1725		–	8	1728	10	605		–	13	1489
Polystyrene, colloidal aggregate	2	965		–		–		–		–		–		–		–		–
Polytetraethylene glycol dimethacrylate		–		–		–		–		–		–		–		–	13	1436
Polytetrafluoroethylene	2	967		–	8	1730	8	1732	8	1734	8	1736	10	608		–	13	1443
Polytetrafluoroethylene, Halon G-80		–		–		–		–		–		–		–		–	13	1445
Polytetrafluoroethylene, stretched PTFE, film		–		–		–		–		–		–		–		–	13	1446
Polythene, PM-1, vinyl polymer		–		–		–		–		–		–		–		–	13	1429
Poly(triallyl citrate)		–		–		–		–		–		–		–		–	13	1392
Polytrifluorochloroethylene	2	970		–		–		–		–		–		–		–	13	1442
Polytrifluorochloroethylene, Kel-F	2	970		–		–		–		–		–		–		–		–
Polytrifluoroethylene		–		–		–		–		–		–	10	611		–		–
Polyurethane	2	982		–		–		–		–		–	10	612		–	13	1494
Polyvinyl chloride	2	953		–		–	8	1742		–	8	1746	10	613		–		–
Poly(vinyl cyclohexene dioxide)		–		–		–		–		–		–		–		–	13	1496
Poly(vinyl toluene)		–		–		–		–		–		–		–		–	13	1497
Polyvinylidene chloride		–		–		–		–		–	8	1746		–		–		–
Pollucite		–		–		–		–		–		–		–		–	13	709
Polonium, Po		–		–		–		–		–		–		–		–	12	270
Porcelain, 576	2	937		–		–		–		–		–		–		–		–
Porcelain, alumina	2	937		–		–		–		–		–		–		–		–
Porcelain, electrical	2	937		–		–		–		–		–		–		–		–
Porcelain, high zircon	2	937		–		–		–		–		–		–		–		–
Porcelain, magnesium titanate	2	937		–		–		–		–		–		–		–		–
Porcelain, wet process	2	937		–		–		–		–		–		–		–		–
Poroloy		–		–	7	1226		–	7	1303		–		–		–		–
Potassium, K	1	274	4	171		–		–		–		–	10	144		–	12	271

Substance Name	Thermal Conductivity		Specif. Heat		Thermal Radiative Properties								Thermal Diffusivity		Viscosity		Thermal Expansion	
					Emissivity		Reflectivity		Absorptivity		Transmissiv.							
	V.	Page	V.	Page	V.	Page	V.	Page	V.	Page	V.	Page	V.	Page	V.	Page	V.	Page
Potassium alloy, K + Na	1	748	4	428		–		–		–		–	10	254		–		–
Potassium alum		–		–		–		–		–	8	1702		–		–		–
Potassium aluminum silicates:																		
KAlSi$_2$O$_6$		–		–		–		–		–		–		–		–	13	727
KAl$_3$Si$_3$O$_{11}$		–	5	1540		–		–		–		–		–		–		–
KAl$_3$Si$_3$O$_{11}$·H$_2$O		–	5	1543		–		–		–		–		–		–		–
Potassium aluminum sulfate, KAl(SO$_4$)$_2$		–	5	1212		–		–		–		–		–		–		–
Potassium aluminum sulfate hydrate, KAl(SO$_4$)$_2$·12H$_2$		–	5	1215		–		–		–		–		–		–		–
Potassium bromide, KBr	2	566	5	765	8	749 751 753		–	8	759		–		–		–	13	810
Potassium bromide + potassium chloride, mixture	2	779		–		–		–		–		–		–		–	13	830
Potassium bromide + rubidium bromide, mixture		–		–		–		–		–		–		–		–	13	838
Potassium carbonate, K$_2$CO$_3$		–	5	1124		–		–		–		–		–		–		–
Potassium chloride, KCl	2	613	5	872	8	862	8	864		·	8	873		–		–	13	982
Potassium chloride + potassium bromide, mixture	2	782		–		–		–		–		–		–		–		–
Potassium chloride + rubidium chloride, mixture		–		–		–		–		·		–		–		–	13	1012
Potassium chrome alum salt	2	689		–		–		–		–		–		–		–		–
Potassium chromium sulfate, KCr(SO$_4$)$_2$·12H$_2$O	2	688		–		–		–		–		–		–		–		–
Potassium cobalt fluoride, KCoF$_3$		–		–		–	8	988		–		–		–		–		–
Potassium deuterium arsenate, KD$_2$AsO$_4$		–		–		–		–		–		–		–		–	13	620
Potassium deuterium phosphate, KD$_2$PO$_4$	2	680		–		–		–		–		–		–		–	13	690
Potassium dihydrogen phosphate, KDP		–				–		–		–	8	604		–		–		–
Potassium dihydrogen arsenate, KH$_2$AsO$_4$	2	785		–		–		–		–		–		–		–	13	623
Potassium dioxide, KO$_2$		–	5	184		–		–		–		–		–		–		–
Potassium fluoride, KF		–	5	979		–	8	992		–		–		–		–	13	1056
Potassium germanium oxides:																		
K$_2$O·2GeO$_2$		–		–		–		–		–		–		–		–	13	498
K$_2$O·7GeO$_2$		–		–		–		–		–		–		–		–	13	498
3K$_2$O·11GeO$_2$		–		–		–		–		–		–		–		–	13	498
Potassium hydrogen fluoride, KHF$_2$		–	5	982		–		–		–		–		–		–		–
Potassium hydrogen phosphate, KH$_2$PO$_4$	2	684		–		–		–		–	8	604		–		–	13	677
Potassium hydrogen selenite, KH$_3$(SeO$_3$)$_2$		–		–		–		–		–		–		–		–	13	694
Potassium hydrogen sulfate, KHSO$_4$	2	691		–		–		–		–		–		–		–		–
Potassium iodide, KI		–	5	500		–	8	1005		–	8	1010		–		–	13	1102
Potassium magnesium fluoride, KMgF$_3$		–		–		–	8	975		–	8	977		–		–		–
Potassium manganese fluoride, KMnF$_3$		–		–		–	8	988		–		–		–		–		–
Potassium nickel fluoride, KNiF$_3$		–		–		–	8	979		–	8	981		–		–		–
Potassium niobium oxide, K$_2$O·Nb$_2$O$_5$		–		–		–		–		–		–		–		–	13	532
Potassium nitrate, KNO$_3$	2	647	5	1145		–		–		–		–		–		–		–

Substance Name	Thermal Conduc- tivity		Specif. Heat		Thermal Radiative Properties								Thermal Diffu- sivity		Visco- sity		Thermal Expan- sion	
					Emis- sivity		Reflec- tivity		Absorp- tivity		Trans- missiv.							
	V.	Page	V.	Page	V.	Page	V.	Page	V.	Page	V.	Page	V.	Page	V.	Page	V.	Page
Potassium oxide + silicon oxide + ΣXi, mixture	2	507	–		–		–		–		–		–		–		–	
Potassium rhodanide, KSCN	2	788	–		–		–		–		–		–		–		–	
Potassium silicon fluoride, K_2SiF_6	–		–		–		8	992	–		–		–		–		–	
Potassium sulfate, K_2SO_4	–		5	1209	–		–		–		–		–		–		13	736
Potassium sulfocyanide, KSCN	2	788	–		–		–		–		–		–		–		–	
Potassium superoxide, KO_2	–		5	184	–		–		–		–		–		–		–	
Potassium tantalum oxide, $K_2O \cdot Ta_2O_5$	–		–		–		8	633	–		–		–		–		13	544
Potassium thiocyanate, KSCN	2	788	–		–		–		–		–		–		–		–	
Potassium zinc fluoride, $KZnF_3$	–		–		–		–		–		8	990	–		–		–	
Potato	–		–		–		–		–		–		10	643	–		–	
Potato starch	–		–		–		–		–		–		10	644	–		–	
Praseodymium, Pr	1	281	4	177	–		–		–		–		10	147	–		12	272
Praseodymium boride, PrB_6	–		–		8	723	–		–		–		–		–		–	
Praseodymium carbide, PrC_2	–		–		–		–		–		–		–		–		13	936
Praseodymium oxides:																		
Pr_2O_3	–		–		–		–		–		–		–		–		13	336
Pr_6O_{11}	–		5	187	–		–		–		–		–		–		13	339
Praseodymium-ruthenium intermetallic compound, $PrRu_2$	–		–		–		–		–		–		–		–		12	610 612 613
Praseodymium selenide, Pr_2Se_3	–		–		–		–		–		–		–		–		13	1192
Praseodymium sulfides:																		
PrS	–		–		–		–		–		–		–		–		13	1240
Pr_2S_3	–		–		8	1232	–		–		–		–		–		13	1240
Praseodymium-tin intermetallic compound, $PrSn_3$	–		–		–		–		–		–		–		–		12	611 612 614
Promethium	1	285	–		–		–		–		–		–		–		–	
Propadiene	–		6s	75	–		–		–		–		–		–		–	
Propane	3	240	6	279	–		–		–		–		–		11	208	–	
1,2-Propanediol	–		6s	76	–		–		–		–		–		–		–	
1-Propandl	–		6s	77	–		–		–		–		–		–		–	
2-Propanol	–		6s	79	–		–		–		–		–		–		–	
2-Propanol, IPA	–		6s	79	–		–		–		–		–		–		–	
2-propanone	3	129	6	113	–		–		–		–		–		–		–	
2-Propen-1-ol	–		6s	1	–		–		–		–		–		–		–	
Propenyl alcohol	–		6s	1	–		–		–		–		–		–		–	
Propine	–		6s	82	–		–		–		–		–		–		–	
Propionic ester	–		6s	38	–		–		–		–		–		–		–	
1-Propoxypropane	–		6s	81	–		–		–		–		–		–		–	
Propyl acetate	–		6s	80	–		–		–		–		–		–		–	
Propyl acetylene	–		6s	73	–		–		–		–		–		–		–	
Propyl alcohol	–		6s	77	–		–		–		–		–		–		–	

Substance Name	Thermal Conductivity		Specif. Heat		Thermal Radiative Properties								Thermal Diffusivity		Viscosity		Thermal Expansion	
					Emissivity		Reflectivity		Absorptivity		Transmissiv.							
	V.	Page	V.	Page	V.	Page	V.	Page	V.	Page	V.	Page	V.	Page	V.	Page	V.	Page
sec-Propyl alcohol		–	6s	79		–		–		–		–		–		–		–
Propyl benzene		–	6s	80		–		–		–		–		–		–		–
Propyl bromide		–	6s	5		–		–		–		–		–		–		–
Propyl carbinol		–	6s	6		–		–		–		–		–		–		–
Propyl chloride		–	6s	22		–		–		–		–		–		–		–
Propyl ethanoate		–	6s	80		–		–		–		–		–		–		–
Propyl ether		–	6s	81		–		–		–		–		–		–		–
Propyl ethylene		–	6s	73		–		–		–		–		–		–		–
Propylene		–		–		–		–		–		–		–	11	213		–
Propylene bromide		–	6s	27		–		–		–		–		–		–		–
Propylene chloride		–	6s	29		–		–		–		–		–		–		–
Propylene dibromide		–	6s	27		–		–		–		–		–		–		–
Propylene dichloride		–	6s	29		–		–		–		–		–		–		–
Propylene glycol		–	6s	76		–		–		–		–		–		–		–
Propyne		–	6s	82		–		–		–		–		–		–		–
Protactinium, Pa		–		–		–		–		–		–		–		–	12	277
Protenstatite		–		–		–		–		–		–		–		–	13	717
Prune juice		–		–		–		–		–		–	10	640		–		–
Prussic acid		–	6s	46		–		–		–		–		–		–		–
Prussite		–	6s	24		–		–		–		–		–		–		–
Pseudobutylene		–	6s	9		–		–		–		–		–		–		–
cis-Pseudobutylene		–	6s	9		–		–		–		–		–		–		–
trans-Pseudobutylene		–	6s	10		–		–		–		–		–		–		–
Pseudocumene		–	6s	93		–		–		–		–		–		–		–
Pseudocumol		–	6s	93		–		–		–		–		–		–		–
Pyrex 7740, borosilicate glass		–		–		–		–		–		–		–		–	13	1356
Pyridine		–	6s	83		–		–		–		–		–		–		–
Pyroacetic acid	3	129		–		–		–		–		–		–		–		–
Pyroacetic ether		–	6	113		–		–		–		–		–		–		–
Pyroacatechin		–	6s	83		–		–		–		–		–		–		–
Pyrocatechnic acid		–	6s	83		–		–		–		–		–		–		–
Pyrocatechol		–	6s	83		–		–		–		–		–		–		–
Pyroceram		–	5	1237		–		–		–		–		–		–		–
Pyroceram 9606	2	940	5	1237	8	1595 1597 1599 1601	8	1603	8	1606		–	10	582 583		–		–
Pyroceram 9608		–	5	1237	8	1595 1597 1599 1601	8	1603	8	1606	8	1608	10	582		–		–
Pyroceram brand glass-ceramic	2	939		–		–		–		–		–	10	582 583		–		–
Pyroceram ceramics		–		–		–		–		–		–		–		–	13	1287
Pyrrolylene		–	6s	5		–		–		–		–		–		–		–
Quartz	2	174	5	207		–		–		–		–	10	397		–	13	352

Substance Name	Thermal Conductivity		Specif. Heat		Thermal Radiative Properties								Thermal Diffusivity		Viscosity		Thermal Expansion	
					Emissivity		Reflectivity		Absorptivity		Transmissiv.							
	V.	Page	V.	Page	V.	Page	V.	Page	V.	Page	V.	Page	V.	Page	V.	Page	V.	Page
Quartz, crystal		–	5	207	8	371 372	8	374 385 386	8	389	8	391		–		–		–
Quartz, fiber Dyna	2	1144		–		–		–		–		–		–		–		–
Quartz, fused		–		–	8	403 405	8	409 413	8	415	8	417 426		–		–	13	360
Quartz, fused, G.E. 106		–		–		–		–		–	8	421		–		–		–
Quartz glass	2	187 188 923 924	5	202		–		–		–		–		–		–		–
Quartz sand	2	834 835 836 837		–		–		–		–		–		–		–		–
Quick silver	1	212	4	131		–		–		–		–		–		–		–
Quinol		–	6s	53		–		–		–		–		–		–		–
Quinone		–	6s	2		–		–		–		–		–		–		–
Radium		–		–		–		–		–		–	10	148		–		–
Radon	3	84		–		–		–		–		–		–		–		–
Rare earth boride		–		–	8	722		–		–	8	726		–		–		–
Reflector plate, Alcoa No. 2		–		–	7	4 5		–	7	42 43		–		–		–		–
Refractories, aluminosilicate		–		–		–		–		–		–	10	564		–		–
Refractory materials		–		–		–		–		–		–	10	563		–		–
Refrax	2	586		–		–		–		–		–		–		–		–
Refrigerants:																		
R-10, carbon tetrachloride		–		–		–		–		–		–		–	11	129		–
R-11, trichlorofluoromethane		–		–		–		–		–		–		–	11	220		–
R-12, dichlorodifluoromethane		–		–		–		–		–		–		–	11	150		–
R-13, chlorotrifluorumethane		–		–		–		–		–		–		–	11	145		–
R-13B1, bromotrifluoromethane		–		–		–		–		–		–		–	11	104		–
R-14, carbon tetrafluoride		–		–		–		–		–		–		–	11	131		–
R-20, chloroform		–		–		–		–		–		–		–	11	138		–
R-21, dichlorofluoromethane		–		–		–		–		–		–		–	11	155		–
R-22, chlorodifluoromethane		–		–		–		–		–		–		–	11	133		–
R-23, trifluoromethane		–		–		–		–		–		–		–	11	230		–
R-40, methyl chloride		–		–		–		–		–		–		–	11	194		–
R-50, methane		–		–		–		–		–		–		–	11	186		–
R-113, trichlorotrifluoroethane		–		–		–		–		–		–		–	11	225		–
R-114, dichlorotetrafluoroethane		–		–		–		–		–		–		–	11	160		–
R-115, chloropentafluoroethane		–		–		–		–		–		·		–	11	140		–
R-152A, 1,1-difluoroethane		–		–		–		–		–		–		–	11	165		–
R-170, ethane		–		–		–		–		–		–		–	11	167		–
R-290, propane		–		–		–		–		–		–		–	11	208		–
R-C318, octafluorocyclobutane		–		–		–		–		–		–		–	11	199		–
R-500, R-12-R-152A		–		–		–		–		–		–		–	11	553		–

Substance Name	Thermal Conductivity		Specif. Heat		Thermal Radiative Properties								Thermal Diffusivity		Viscosity		Thermal Expansion	
					Emissivity		Reflectivity		Absorptivity		Transmissiv.							
	V.	Page	V.	Page	V.	Page	V.	Page	V.	Page	V.	Page	V.	Page	V.	Page	V.	Page
Refrigerants: (continued)																		
R-502, R-12-R-115	–		–		–		–		–		–		–		11	558	–	
R-503, R-13-R-23	–		–		–		–		–		–		–		11	563	–	
R-504, CH_2F_2 + R-115	–		–		–		–		–		–		–		11	565	–	
R-600, *n*-butane	–		–		–		–		–		–		–		11	114	–	
R-600A, *i*-butane	–		–		–		–		–		–		–		11	109	–	
R-610, ethyl ether	–		–		–		–		–		–		–		11	180	–	
R-702, hydrogen	–		–		–		–		–		–		–		11	24	–	
R-704, helium	–		–		–		–		–		–		–		11	18	–	
R-704A, deuterium	–		–		–		–		–		–		–		11	13	–	
R-717, ammonia	–		–		–		–		–		–		–		11	68	–	
R-718, water	–		–		–		–		–		–		–		11	94	–	
R-720, neon	–		–		–		–		–		–		–		11	41	–	
R-728, nitrogen	–		–		–		–		–		–		–		11	48	–	
R-728A, carbon monoxide	–		–		–		–		–		–		–		11	125	–	
R-729, air	–		–		–		–		–		–		–		11	608	–	
R-730, nitric oxide	–		–		–		–		–		–		–		11	82	–	
R-732, oxygen	–		–		–		–		–		–		–		11	56	–	
R-734, hydrogen sulfide	–		–		–		–		–		–		–		11	80	–	
R-736, hydrogen chloride	–		–		–		–		–		–		–		11	76	–	
R-738, fluorine	–		–		–		–		–		–		–		11	16	–	
R-740, argon	–		–		–		–		–		–		–		11	2	–	
R-744, carbon dioxide	–		–		–		–		–		–		–		11	119	–	
R-744A, nitrous oxide	–		–		–		–		–		–		–		11	87	–	
R-746, nitrogen peroxide	–		–		–		–		–		–		–		11	85	–	
R-764, sulfur dioxide	–		–		–		–		–		–		–		11	91	–	
R-768, boron trifluoride	–		–		–		–		–		–		–		11	74	–	
R-771, chlorine	–		–		–		–		–		–		–		11	11	–	
R-784, krypton	–		–		–		–		–		–		–		11	37	–	
R-1150, ethylene	–		–		–		–		–		–		–		11	174	–	
R-1270, propylene	–		–		–		–		–		–		–		11	213	–	
Resin	–		–		–		–		–		–		10	614	–		–	
Resin, epoxy	–		–		–		–		–		–		–		–		13	1502
Resin, polyester	–		–		–		–		–		–		–		–		13	1511
Resin, phenoxy	–		–		–		–		–		–		–		–		13	1512
Resin, phenolic	–		–		–		–		–		–		10	615	–		13	1509
Resin, vinyl	–		–		–		8	1741	8	1743	8	1745	–		–		–	
Resin cured butyl pseudo balsa	2	983	–		–		–		–		–		–		–		–	
Resorcin	–		6s	83	–		–		–		–		–		–		–	
Resorcinol	–		6s	83	–		–		–		–		–		–		–	
Rhenium, Re	1	288	4	181	7	559 562 565 568	–		–		–		10	149	–		12	280

Substance Name	Thermal Conductivity		Specif. Heat		Thermal Radiative Properties								Thermal Diffusivity		Viscosity		Thermal Expansion	
					Emissivity		Reflectivity		Absorptivity		Transmissiv.							
	V.	Page	V.	Page	V.	Page	V.	Page	V.	Page	V.	Page	V.	Page	V.	Page	V.	Page
Rhenium arsenide, Re₃As₇	1	1330	–		–		–		–		–		–		–		–	
Rhenium chloride, ReCl₃	–		5	878	–		–		–		–		–		–		–	
Rhenium–germanium intermettalic compounds:																		
RhGe	1	1331	–		–		–		–		–		–		–		–	
RhGe₂	1	1331	–		–		–		–		–		–		–		–	
Rhenium oxide, ReO₃	–		–		–		8	546	–		–		–		–		–	
Rhenium selenide, ReSe₂	1	1332	–		–		–		–		–		–		–		–	
Rhenium silicides:																		
ReSi	–		–		–		–		–		–		–		–		13	1213
ReSi₂	–		–		8	1173	–		–		–		–		–		13	1213
Rhodium, Rh	1	292	4	184	7	571 573 576 579	7	581 584	7	587 589	–		10	152	–		12	285
Rhodium alloys:																		
Rh + Fe	–		–		–		–		–		–		–		–		12	864
Rh + Mo	–		–		–		–		–		–		–		–		12	904
Rh + Pt	–		–		–		–		–		–		–		–		12	967
Rhyolite tuff	–		–		–		8	1682	–		–		–		–		–	
Rock	2	828	–		–		8	1680	–		–		–		–		–	
Rock, minerals	–		–		–		–		–		–		10	545	–		–	
Rock, Winchester crushed trap	2	829 830	–		–		–		–		–		–		–		–	
Rock cork	2	1146	–		–		–		–		–		–		–		–	
Rock salt, natural	–		–		–		–		–		–		–		–		13	1003
Rock salt, synthetic	–		–		–		–		–		–		–		–		13	1003
Rock wool	2	1148	–		–		–		–		–		–		–		–	
Rose metal	1	939	–		–		–		–		–		–		–		–	
Rubber, acrylate	2	982	–		–		–		–		–		–		–		–	
Rubber, acrylic	2	982	–		–		–		–		–		–		–		–	
Rubber, adiprene	2	982	–		–		–		–		–		–		–		–	
Rubber, butaprene E	2	982	–		–		–		–		–		–		–		–	
Rubber, carboxy nitrile	2	982	–		–		–		–		–		–		–		–	
Rubber, chloroprene	2	983	–		–		–		–		–		–		–		–	
Rubber, dibenzo GMF-cured butyl pseudo balsa	2	983	–		–		–		–		–		–		–		–	
Rubber, Ebonite	2	971	–		–		–		–		–		10	617	–		–	
Rubber, elastomer	2	974	–		–		–		–		–		–		–		–	
Rubber, epoxy	–		–		–		–		–		–		–		–		13	1520
Rubber, EPR	–		–		–		–		–		–		–		–		13	1520
Rubber, foam buna-N	2	981	–		–		–		–		–		–		–		–	
Rubber, halveg elastomer	–		–		–		–		–		–		–		–		13	1520
Rubber, hevea	2	983	–		–		–		–		–		–		–		–	
Rubber, Hycar 4021	–		–		–		–		–		–		–		–		13	1423

Substance Name	Thermal Conduc-tivity		Specif. Heat		Thermal Radiative Properties										Thermal Diffu-sivity		Visco-sity		Thermal Expan-sion	
					Emis-sivity		Reflec-tivity		Absorp-tivity		Trans-missiv.									
	V.	Page	V.	Page	V.	Page	V.	Page	V.	Page	V.	Page	V.	Page	V.	Page	V.	Page
Rubber, hypalon 5-20		–		–		–		–		–		–		–		–	13	1434
Rubber, hypalon S2	2	983		–		–		–		–		–		–		–		–
Rubber, Kel-F 3700	2	983		–		–		–		–		–		–		–		–
Rubber, methacrylate	2	983		–		–		–		–		–		–		–		–
Rubber, miscellaneous polymer		–		–		–		–		–		–		–		–	13	1520
Rubber, neoprene		–		–		–		–		–		–		–		–	13	1514
Rubber, nitrile	2	982		–		–		–		–		–		–		–		–
Rubber, polysulfide	2	982		–		–		–		–		–		–		–		–
Rubber, PR 19-10		–		–		–	8	1740		–		–		–		–		–
Rubber, RTV-77		–		–		–	8	1740		–		–		–		–		–
Rubber, rubatex R203-H	2	981		–		–		–		–		–		–		–		–
Rubber, rubatex	2	981		–		–		–		–		–		–		–		–
Rubber, silicone	2	983		–		–		–		–		–	10	617		–	13	1517
Rubber, silicone silastic 160		–		–		–		–		–		–		–		–	13	1518
Rubber, silicone silastic 400		–		–		–		–		–		–		–		–	13	1518
Rubber, silicone silastic X30028		–		–		–		–		–		–		–		–	13	1518
Rubber, styrene	2	977		–		–		–		–		–		–		–		–
Rubber, Thiokol ST	2	982		–		–		–		–		–		–		–		–
Rubber, Viton	2	983		–		–		–		–		–		–		–		–
Rubidium, Rb	1	296	4	187		–		–		–		–	10	153		–		–
Rubidium alloy, Rb + Cs	1	751		–		–		–		–		–		–		–		–
Rubidium aluminum silicate, $RbAlSi_2O_6$		–		–		–		–		–		–		–		–	13	728
Rubidium bromide, RbBr		–	5	769	8	764	8	766		–	8	768		–		–	13	816
Rubidium bromide + rubidium chloride, RbBr + RbCl, mixtur		–		–		–		–		–		–		–		–	13	834
Rubidium chloride, RbCl		–		–		–	8	906		–	8	908		–		–	13	990
Rubidium dideuterium arsenate, RbD_2AsO_4		–		–		–		–		–		–		–		–	13	627
Rubidium fluoride, RbF		–	5	985		–		–		–		–		–		–	13	1076
Rubidium dihydrogen arsenate, RbH_2AsO_4		–		–		–		–		–		–		–		–	13	630
Rubidium hydrogen fluoride, $RbHF_2$		–	5	988		–		–		–		–		–		–		–
Rubidium dihydrogen orthophosphate, RbH_2PO_4		–		–		–		–		–		–		–		–	13	690
Rubidium iodide, RbI		–	5	503		–	8	1014		–	8	1018		–		–	13	1109
Rubidium manganese fluoride, $RbMnF_3$		–		–		–	8	983		–	8	985		–		–	13	1076
Rubidium nitrate, $RbNO_3$		–		–		–		–		–		–		–		–	13	671
Rubidium sulfate, Rb_2SO_4		–		–		–		–		–		–		–		–	13	739
Ruby		–		–		–	8	174		–	8	176		–		–		–
Ruby, spinel, natural	2	284		–		–		–		–		–		–		–		–
Ruthenium, Ru	1	300	4	190	7	591		–		–		–	10	154		–	12	290
Ruthenium oxide, RuO_2		–		–		–		–		–		–		–		–	12	341
Rutile	2	203	5	246		–	8	464 465		–		–		–		–	13	394
Rutile, nigrine	2	203		–		–		–		–		–		–		–		–
SAE 1010, steel	1	1183	4	647		–		–		–		–		–		–	12	842 1165 1167

Substance Name	Thermal Conductivity		Specif. Heat		Thermal Radiative Properties										Thermal Diffusivity		Viscosity		Thermal Expansion	
					Emissivity		Reflectivity		Absorptivity		Transmissiv.									
	V.	Page	V.	Page	V.	Page	V.	Page	V.	Page	V.	Page	V.	Page	V.	Page	V.	Page
SAE 1015, steel	1	1186	–		–		–		–		–		–		–		–	
SAE 1020, steel	1	1183	–		–		–		–		–		10	354	–		12	1165 1167
SAE 1095, steel	1	1114	–		–		–		–		–		–		–		–	
SAE 4130, steel	1	1153	–		–		–		–		–		10	339	–		–	
SAE 4140, steel	1	1155	–		–		–		–		–		–		–		–	
SAE 4340, steel	1	1213 1214	–		–		–		–		–		10	363 364	–		–	
SAE bearing alloy 10, tin alloy	1	1070	–		–		–		–		–		–		–		–	
SAE bearing alloy 11, tin alloy	1	1070	–		–		–		–		–		–		–		–	
SAE bearing alloy 12, lead alloy	1	991	–		–		–		–		–		–		–		–	
SAE bearing alloy 40, copper alloy	1	976	–		–		–		–		–		–		–		–	
SAE bearing alloy 62, copper alloy	1	976	–		–		–		–		–		–		–		–	
SAE bearing alloy 64, copper alloy	1	976	–		–		–		–		–		–		–		–	
SAE bearing alloy 66, copper alloy	1	962	–		–		–		–		–		–		–		–	
Salt, gnome	2	832	–		–		–		–		–		–		–		–	
Salt, pool		–			–		8	1660	–		–		–		–		–	
Samaria		–	5	193	–		–		–		–		–		–		–	
Samarium, Sm	1	305	4	193	–		–		–		–		10	155	–		–	
Samarium boride, SmB_2		–		–	8	723	–		–		–		–		–		13	797
Samarium carbide, SmC_2		–		–	–		–		–		–		–		–		13	936
Samarium oxide, Sm_2O_3		–	5	193	8	352 354 356 358	8	360	–		–		–		–		13	344
Samarium selenide, Sm_2Se_3		–		–	–		–		–		–		–		–		13	1192
Samarium silicate		–		–	–		–		–		8	622	–		–		–	
Samarium-silver intermetallic compound, $SmAg_3$					–		–		–		–		–		–		12	615
Samarium sulfide, Sm_2S_3		–		–	–		–		–		–		–		–		13	1240
Sand		–		–	–		8	1678	–		–		–		–		–	
Sand, lowell	2	834 835	–		–		–		–		–		–		–		–	
Sand, silica	2	837	–		–		–		–		–		–		–		–	
Sandstones:																		
Berea	2	841 842	–		–		–		–		–		–		–		–	
Berkeley	2	841 842	–		–		–		–		–		–		–		–	
Sandstone	2	840	–		–		–		–		–		–		–		–	
St. Peters	2	841	–		–		–		–		–		–		–		–	
Teapot	2	842	–		–		–		–		–		–		–		–	
Tensleep	2	841 842	–		–		–		–		–		–		–		–	
Tripolite	2	842	–		–		–		–		–		–		–		–	
Santowax R	2	1005	–		–		–		–		–		–		–		–	
Sapphire	2	93	–		8	179 181 183	8	187	–		8	190	–		–		13	178 180 182 184

Substance Name	Thermal Conductivity		Specif. Heat		Thermal Radiative Properties								Thermal Diffusivity		Viscosity		Thermal Expansion	
					Emissivity		Reflectivity		Absorptivity		Transmissiv.							
	V.	Page	V.	Page	V.	Page	V.	Page	V.	Page	V.	Page	V.	Page	V.	Page	V.	Page
Sapphire, Linde synthetic	2	94	–		–		–		–		–		–		–		–	
Sapphire, synethetic	2	95	–		–		–		–		–		–		–		–	
Scandia		–	5	196	–		–		–		–		–		–		–	
Scandium, Sc	1	309	4	198	–		–		–		–		10	156	–		12	295
Scandium boride, ScB₂		–		–	8	732	–		–		–		–		–		13	797
Scandium oxide, Sc₂O₃		–		–	–		–		–		–		–		–		13	347
Scandium nitride, ScN		–		–	8	1087	–		–		8	1090	–		–		13	1162
Scolecite		–		–	–		–		–		8	1694	–		–		–	
Scorched earth + refractory clay + dolomite, mixture		–		–	–		–		–		–		10	433	–		–	
Sea-weed, pressed powder	2	1128		–	–		–		–		–		–		–		–	
Selenides, miscellaneous		–		–	–		8	1129	–		8	1132	–		–		–	
Selenite		–		–	–		–		–		8	630	–		–		–	
Selenium, Se	1	313	4	201	–		8	80	–		8	84	10	157	–		13	149
Selenium alloys:																		
Se + Br	1	754	–		–		–		–		–		–		–		–	
Se + Cd	1	755	–		–		–		–		–		–		–		–	
Se + Cl	1	756	–		–		–		–		–		–		–		–	
Se + I	1	757	–		–		–		–		–		–		–		–	
Se + Tl	1	758	–		–		–		–		–		–		–		–	
Serpentine		–		–	–		8	1689	–		8	1694	–		–		–	
Silane		–	6s	83	–		–		–		–		–		–		–	
Silica	2	185	5	202 207 210 213	–		–		–		–		–		–		–	
Silica, crystal		–		–	8	371 372	8	374 385 387	8	389	8	391	–		–		–	
Silica, fused		–		–	8	403 405	8	409 413	8	415	8	417 426	10	399	–		13	360
Silica gel		–		–	–		–		–		8	422	–		–		–	
Silicane		–	6s	83	–		–		–		–		–		–		–	
Silicates, miscellaneous		–		–	–		8	617	–		8	621	–		–		–	
Silicides, miscellaneous		–		–	8	1172	8	1175	–		–		–		–		–	
Silicochloroform		–	6s	92	–		–		–		–		–		–		–	
Silicon, Si	1	326	4	204	8	88 89 90	8	97 106	8	108	8	110	10	160	–		13	154
Silicon, oxidized		–		–	–		9	1316 1318	–		9	1320	–		–		–	
Silicon alloys:																		
Si + Fe	1	764	–		–		–		–		–		–		–		–	
Si + Fe, Russian, ferrosilicon	1	765	–		–		–		–		–		–		–		–	
Si + Ge	1	761	–		–		–		–		–		–		–		–	

Substance Name	Thermal Conductivity		Specif. Heat		Thermal Radiative Properties								Thermal Diffusivity		Viscosity		Thermal Expansion	
					Emissivity		Reflectivity		Absorptivity		Transmissiv.							
	V.	Page	V.	Page	V.	Page	V.	Page	V.	Page	V.	Page	V.	Page	V.	Page	V.	Page
Silicon borides:																		
SiB₄		–		–		–		–		–		–		–		–	13	797
SiB₆		–		–		–		–		–		–		–		–	13	797
Silicon carbides:																		
SiC	2	585	5	448		–		–		–		–	10	477		–	13	873
SiC brick, refrax	2	586		–	8	791 792 796 798	8	802	8	808 810		–				–		–
Crystolon SiC	2	586		–		–		–		–		–		–		–		–
KT		–		–	8	794		–		–		–		–		–		–
Refractory	2	586		–		–		–		–		–		–		–		–
Silicon carbide + silicon, mixture	2	718		–		–		–		–		–		–		–		–
Silicon carbide + silicon dioxide, mixture	2	553		–		–		–		–		–		–		–		–
Silicon carbide + silicon dioxide + ΣXi, mixture	2	554		–		–		–		–		–		–		–		–
Silicon carbide + zirconium boride, mixture		–		–		–		–		–		–	10	536		–		–
Silicon chloride, SiCl₄		–	5 6s	881 83		–		–		–		–		–		–		–
Silicon dioxide + aluminum, cermet		–		–		–	8	1428		–		–		–		–		–
Silicon dioxide + zirconium oxide + ΣXi, mixture	2	534		–		–		–		–		–		–		–		–
Silicon fluoride, SiF₄		–	6s 5	84 991		–		–		–		–		–		–		–
Silicon hydride		–	6s	83		–		–		–		–		–		–		–
Silicon nitride, Si₃N₄	2	662	5	1087	8	1062 1065 1067	8	1069		–		–		–		–	13	1140
Silicon oxides:																		
SiO		–		–	8	362	8	365		–	8	367		–		–		–
SiO₂		–		–		–		–		–	9	1073		–		–		–
Cristobalite		–	5	210		–		–		–		–		–		–		–
Crystalline	2	174		–	8	371 372	8	374 385 387	8	389	8	391	10	396		–	13	350
Domestic, USA	2	175		–		–		–		–		–		–		–		–
Foamed fused silica	2	184		–		–		–		–		–		–		–		–
Fused	2	183		–	8	403 405	8	409 413	8	415	8	417 426	10	399		–	13	358
Linde silica	2	184		–		–		–		–		–		–		–		–
Quartz		–	5	202		–		–		–		–		–		–		–
Quartz glass	2	187 188		–		–		–		–		–		–		–		–
Silica gel	2	185		–		–		–		–		–		–		–		–
Silica refractory brick	2	185		–		–		–		–		–		–		–		–
Slip 10	2	189		–		–		–		–		–		–		–		–
Slip 18	2	188		–		–		–		–		–		–		–		–
Slip cast fused silica	2	184		–		–		–		–		–		–		–		–
Star-brand brick	2	185		–		–		–		–		–		–		–		–

Substance Name	Thermal Conduc-tivity		Specif. Heat		Thermal Radiative Properties										Thermal Diffu-sivity		Visco-sity		Thermal Expan-sion	
					Emis-sivity		Reflec-tivity		Absorp-tivity		Trans-missiv.									
	V.	Page	V.	Page	V.	Page	V.	Page	V.	Page	V.	Page	V.	Page	V.	Page	V.	Page	V.	Page
Silicon oxides: (continued)																				
Vitreous	2	184 185 187		–		–		–		–		–		–		–		–		–
Tridymite		–	5	213		–		–		–		–		–		–		–		–
Silicon oxide + chromium, cermet		–		–		–		–		–	8	1401		–		–		–		–
Silicon oxide + sodium oxide + ΣXi, mixture	2	510		–		–		–		–		–	10	441		–		–		–
Silicon oxide + titanium boride powders		–		–		–	8	1484		–		–		–		–		–		–
Silicone		–		–		–		–		–		–	10	619		–		–		–
Silicone, cellular		–		–		–		–		–		–	10	619		–		–		–
Silicone, Knapic	1	327		–		–		–		–		–		–		–		–		–
Silk fabric	2	1105		–		–		–		–		–		–		–		–		–
Sillimanite	2	454 845	5	1289		–		–		–		–		–		–		–		–
Silon	2	959		–		–		–		–		–		–		–		–		–
Silumin, sodium modified	2	920		–		–		–		–		–		–		–		–		–
Silvan		–	6s	63		–		–		–		–		–		–		–		–
Silver, Ag	1	340	4	208	7	620 623 625 627	7	630 636	7	639 641 643 645 648	7	651	10	164		–	12	298
Silver, 0.6 percent impurities	1	1061		–		–		–		–		–		–		–		–
Silver, electrolytic		–	4	208		–		–		–		–		–		–		–
Silver, Mealtone hammer finish		–		–		–	9	529		–		–		–		–		–
Silver, inquartation		–	4	208		–		–		–		–		–		–		–
Silver alloys:																		
Ag + Al		–		–		–	7	1007 1009		–		–		–		–		–
Ag + Au	1	774		–		–	7	1015		–		–		–		–		–
Ag + Be		–		–		–	7	1012		–		–		–		–		–
Ag + Cd	1	770		–		–		–		–		–		–		–	12	690
Ag + Cu	1	773		–		–	7	1475		–		–		–		–		–
Ag + In	1	777		–		–		–		–		–		–		–	12	826
Ag + Mn	1	783		–		–		–		–		–		–		–	12	893
Ag + Pb	1	780		–		–		–		–		–		–		–		–
Ag + Pd	1	786		–		–		–		–		–		–		–	12	962
Ag + Pt	1	790		–		–		–		–		–		–		–		–
Ag + Sb	1	767		–		–		–		–		–		–		–	12	674
Ag + Si		–		–		–	7	1018		–		–		–		–		–
Ag + Sn	1	791		–		–		–		–		–		–		–	12	976
Ag + Zn	1	792		–		–		–		–		–		–		–		–
Ag + Cd + ΣXi	1	1058		–		–	7	1472		–		–		–		–		–
Ag + Cd + ΣXi, silver solder, easy-flo	1	1059		–		–		–		–		–		–		–		–
Ag + Zn + ΣXi		–		–		–	7	1478		–		–		–		–		–

Substance Name	Thermal Conductivity		Specif. Heat		Thermal Radiative Properties								Thermal Diffusivity		Viscosity		Thermal Expansion	
					Emissivity		Reflectivity		Absorptivity		Transmissiv.							
	V.	Page	V.	Page	V.	Page	V.	Page	V.	Page	V.	Page	V.	Page	V.	Page	V.	Page
Silver-aluminum intermetallic compound, Ag_2Al		–		–		–	8	1352		–		–		–		–		–
Silver-antimony-tellurium intermetallic compound, AgSb	1	1335		–		–		–		–		–		–		–		–
Silver antimony telluride + tin telluride, mixture	1	1410 1411		–		–		–		–		–		–		–		–
Silver antimony telluride, $AgSbTe_2$	1	1335		–		–		–		–		–		–		–		–
Silver bromide, AgBr	2	569		–		–	8	770 773		–	8	775	10	479		–	13	836
Silver bronze	1	579 980		–		–		–		–		–		–		–		–
Silver-cadmium intermetallic compound, AgCd		–		–	.	–	8	1326		–		–		–		–		–
Silver carbonate, Ag_2CO_3		–	5	1127		–		–		–		–		–		–		–
Silver chloride, AgCl	2	620	5	884		–	8	876		–	8	879		–		–	13	995
Silver-copper intermetallic compound, AgCu	1	1338		–		–		–		–		–		–		–		–
Silver iodide, AgI	2	563		–		–	8	1022		–	8	1024		–		–	13	1114
Silver nitrate, $AgNO_3$	2	650		–		–		–		–		–		–		–	13	655
Silver nitrite, $AgNO_2$		–	5	1148		–		–		–		–		–		–		–
Silver oxide, Ag_2O		–	5	199		–		–		–		–		–		–		–
Silver selenide, Ag_2Se	1	1339	5	553		–		–		–		–		–		–		–
Silver selenide, nonstoichiometric			5	556		–		–		–		–		–		–		–
Silver solder, easy-flo	1	1059		–		–		–		–		–		–		–		–
Silver solder, silver alloy easy-flo	1	1059		–		–		–		–		–		–		–		–
Silver sulfide, nonstoichiometric		–	5	705		–		–		–		–		–		–		–
Silver telluride, Ag_2Te	1	1342	5	753		–		–		–		–		–		–		–
Silver telluride, nonstoichiometric			5	756		–		–		–		–		–		–		–
Silver-terbium intermetallic compound, Ag_2Tb		–		–		–		–		–		–		–		–	12	618
Silver thioarsenate, Ag_3AsS_3		–				–		–		–		–		–		–	13	1240
Silver-zinc intermetallic compound, AgZn		–		–		–		–		–		–		–		–	12	619
Skyspar A423		–		–	9	211	9	211	9	252 255 263		–		–		–		–
Slag, mystic	2	1150		–		–		–		–		–		–		–		–
Slag wool	2	1151		–		–		–		–		–		–		–		–
Slate	2	846		–		–		–		–		–		–		–		–
SNAP fuel		–		–		–		–		–		–	10	541		–		–
Snow		–		–		–		–		–		–	10	390		–		–
Soapstone	2	853		–		–		–		–		–		–		–		–
Soda, baking		–	5	1133		–		–		–		–		–		–		–
Sodium, Na	1	349	4 6s	213 53		–		–		–		–	10	167		–	12	310
Sodium, electrolytic		–	4	215		–		–		–		–		–		–		–

Substance Name	Thermal Conductivity		Specif. Heat		Thermal Radiative Properties								Thermal Diffusivity		Viscosity		Thermal Expansion	
					Emissivity		Reflectivity		Absorptivity		Transmissiv.							
	V.	Page	V.	Page	V.	Page	V.	Page	V.	Page	V.	Page	V.	Page	V.	Page	V.	Page
Sodium alloys:																		
Na + K	1	798	4	431		–		–		–		–		–		–		–
Na + Hg	1	795		–		–		–		–		–		–		–		–
Sodium acetate	2	1006		–		–		–		–		–		–		–		–
Sodium aluminum oxide, $Na_2O \cdot Al_2O_3$		–	5	1549		–		–		–		–		–		–		–
Sodium aluminum fluoride, Na_3AlF_6		–	5	997		–		–		–		–		–		–		–
Sodium aluminum silicate, $NaAlSi_3O_8$		–	5	1602		–		–		–		–		–		–	13	728
Sodium bicarbonate, $NaHCO_3$		–	5	1133		–		–		–		–		–		–		–
Sodium borate		–		–	8	582		–		–		–		–		–		–
Sodium bromate, $NaBrO_3$		–		–		–		–		–		–		–		–	13	633
Sodium bromide, $NaBr$		–	5	772		–		–		–		–		–		–	13	821
Sodium calcium silicate, Na_2CaSiO_4		–		–		–		–		–		–		–		–	13	728
Sodium carbonate, Na_2CO_3		–	5	1130		–	8	593		–		–		–		–		–
Sodium chlorate, $NaClO_3$		–		–		–	8	594		–		–		–		–	13	648
Sodium chlorate + sodium nitrate, mixture		–		–		–		–		–		–		–	11	567		–
Sodium chloride, $NaCl$	2	621	5	887	8	881 883	8	885 886 893		–	8	895	10	481		–	13	1000
Sodium ferrite		–	5	1560		–		–		–		–		–		–		–
Sodium fluoride, NaF	2	642	5	994		–	8	963		–	8	966		–		–	13	1060
Sodium fluoride + zirconium tetrafluoride + ΣXi, mixture	2	646		–		–		–		–		–		–		–		–
Sodium hexafluoroaluminate		–	5	997		–		–		–		–		–		–		–
Sodium hydrate, $NaOH$	2	790		–		–		–		–		–		–		–		–
Sodium hydrogen carbonate		–	5	1133		–		–		–		–		–		–		–
Sodium hydrogen fluoride, $NaHF_2$		–	5	1000		–		–		–		–		–		–		–
Sodium hydrogen sulfate, $NaHSO_4$	2	692		–		–		–		–		–		–		–		–
Sodium hydroxide, $NaOH$	2	790		–		–		–		–		–		–		–		–
Sodium hyposulfite, $Na_2S_2O_3 \cdot 5H_2O$	2	693		–		–		–		–		–		–		–		–
Sodium iodate, $NaIO_4$		–		–		–		–		–		–		–		–	13	652
Sodium iodide, NaI		–	5	506		–		–		–		–		–		–	13	1116
Sodium iron dioxide		–	5	1560		–		–		–		–		–		–		–
Sodium lanthanum molybdenum oxide, $Na_2O \cdot La_2O_3 \cdot 4M$		–		–		–		–		–		–		–		–	13	521
Sodium molybdenum oxides:																		
$Na_2O \cdot MoO_3$		–	5	1563		–		–		–		–		–		–		–
$Na_2O \cdot 2MoO_3$		–	5	1566		–		–		–		–		–		–		–
Sodium nickel fluoride, $NaNiF_3$		–		–		–		–		–	8	990		–		–		–
Sodium niobium oxide, $Na_2O \cdot Nb_2O_5$		–		–		–		–		–		–		–		–	13	533
Sodium nitrate, $NaNO_3$	2	651	5	1151		–	8	600		–		–		–		–	13	657
Sodium nitrite, $NaNO_2$		–		–		–		–		–		–		–		–	13	661 662 663 664 665 666 667

Substance Name	Thermal Conductivity		Specif. Heat		Emissivity		Reflectivity		Absorptivity		Transmissiv.		Thermal Diffusivity		Viscosity		Thermal Expansion	
	V.	Page	V.	Page	V.	Page	V.	Page	V.	Page	V.	Page	V.	Page	V.	Page	V.	Page
Sodium oxide, Na_2O		–	5	216		–		–		–		–		–		–		–
Sodium oxide + sodium, cermet	2	721		–		–		–		–		–		–		–		–
	1	1432																
Sodium phosphate		–		–		–	8	608		–		–		–		–		–
Sodium silicates:																		
$\quad Na_2SiO_3$		–	5	1569		–		–		–		–		–		–		–
$\quad Na_2Si_2O_5$		–	5	1572		–		–		–		–		–		–		–
Sodium sulfates:																		
$\quad Na_2SO_4$		–	5	1218		–		–		–		–		–		–	13	740
$\quad Na_2S_2O_3 \cdot 5H_2O$	2	693		–		–		–		–		–		–		–		–
$\quad Na_2SO_4 \cdot 10H_2O$		–	5	1221		–		–		–		–		–		–		–
Sodium tellurate, Na_2TeO_4		–	5	1575		–		–		–		–		–		–		–
Sodium tetrafluoroborate, $NaBF_4$		–		–		–		–		–		–		–		–	13	1076
Sodium-thallium intermetallic compound, NaTl		–		–		–		–		–		–		–		–	12	622
Sodium titanium oxides:																		
$\quad Na_2O \cdot TiO_2$		–	5	1578		–		–		–		–		–		–		–
$\quad Na_2O \cdot 2TiO_2$		–	5	1581		–		–		–		–		–		–		–
$\quad Na_2O \cdot 3TiO_2$		–	5	1584		–		–		–		–		–		–		–
Sodium trideuterium selenite, $NaD_3(SeO_3)_2$		–		–		–		–		–		–		–		–	13	697
Sodium trihydrogen selenite, $NaH_3(SeO_3)_2$		–		–		–		–		–		–		–		–	13	699
Sodium tripolyphosphate		–		–		–	8	608		–		–		–		–		–
Sodium tungsten oxides:																		
$\quad Na_2O \cdot WO_3$		–	5	1587		–	8	666		–		–		–		–	13	592
$\quad Na_2O \cdot 2WO_3$		–	5	1590		–		–		–		–		–		–		–
Sodium vanadium oxides:																		
$\quad Na_2O \cdot V_2O_5$		–	5	1593		–	8	667		–		–		–		–		–
$\quad 2Na_2O \cdot V_2O_5$		–	5	1599		–		–		–		–		–		–		–
$\quad 3Na_2O \cdot V_2O_5$		–	5	1596		–		–		–		–		–		–		–
Soil	2	847		–		–		–		–		–	10	549		–		–
Soil, sandy clay	2	805		–		–		–		–		–		–		–		–
Solar cell, IRC		–		–	8	88 92	8	100		–		–		–		–		–
Solder, Pb + Sn		–	4	446	7	948		–		–		–		–		–		–
Solder, soft	1	840		–		–		–		–		–		–		–		–
Spacemetal		–		–		–		–		–		–	10	552		–		–
Spectrosil		–		–		–		–		–		–		–		–	13	365 366
Spectral kohle 1	2	54		–		–		–		–		–		–		–		–
Spinel	2	284 369 848		–	8	1674	8	576		–	8	578	10	417 419 428		–	13	479
Spinel, natural ruby	2	284		–		–		–		–		–		–		–		–
Spinel, synthetic	2	287		–		–		–		–		–		–		–		–
Spinel firebrick	2	905		–		–		–		–		–		–		–		–

Substance Name	Thermal Conductivity		Specif. Heat		Thermal Radiative Properties								Thermal Diffusivity		Viscosity		Thermal Expansion	
					Emissivity		Reflectivity		Absorptivity		Transmissiv.							
	V.	Page	V.	Page	V.	Page	V.	Page	V.	Page	V.	Page	V.	Page	V.	Page	V.	Page
Spodumene	2	851	–		–		–		–		–		–		–		–	
Spruce	2	1086	–		–		–		–		–		–		–		–	
Stannia	–		5	240	–		–		–		–		–		–		–	
Stannic chloride	–		6s	91	–		–		–		–		–		–		–	
Stannic oxide, SnO_2	2	199	5	240	–		8	451	–		–		–		–		–	
Stannic selenide, $SnSe_2$	1	1352	–		–		–		–		–		–		–		–	
Stannous oxide, SnO	–		5	237	–		–		–		–		–		–		–	
Stannous telluride, SnTe	1	1355	–		–		8	1259	–		8	1264	–		–		–	
Stannum	1	389	–		–		–		–		–		–		–		–	
Starch			–		–		–		–		–		10	644	–		–	
Steatite	2	852	–		–		–		–		–		–		–		–	
Steatite, 10B2	2	853	–		–		–		–		–		–		–		–	
Steatite, 12C2	2	853	–		–		–		–		–		–		–		–	
Steatite, 228	2	853	–		–		–		–		–		–		–		–	
Steatite, cordierite	2	919	–		–		–		–		–		–		–		–	
Steatite, soapstone	2	853	–		–		–		–		–		–		–		–	
Steel	1	1214	–		–		–		–		–		–		–		12	1175 1177
Steel, 19	–		4	687	–		–		–		–		–		–		–	
Steel, 23 D 245	–		–		–		–		–		–		10	342	–		–	
Steel, 23 H 566	–		–		–		–		–		–		10	368	–		–	
Steel, 35 G18	–		–		–		–		–		–		–		–		12	1167 1168
Steel, 40 G5	–		–		–		–		–		–		–		–		12	1169 1170
Steel, 40 N7	–		–		–		–		–		–		–		–		12	1185
Steel, 45 G10	–		–		–		–		–		–		–		–		12	1167
Steel, 2800	–		–		–		–		–		–		–		–		12	1177
Steel, AISI 1010	1	1185	–		–		–		–		–		–		–		–	
Steel, AISI 1010 C	1	1183	–		–		–		–		–		–		–		12	1167
Steel, AISI 1015 C	1	1186	–		–		–		–		–		–		–		–	
Steel, AISI 1018	–		–		–		–		–		–		10	358	–		–	
Steel, AISI 1020 C	1	1183	–		–		–		–		–		–		–		–	
Steel, AISI 1045	–		–		–		–		–		–		10	358	–		–	
Steel, AISI 1095	1	1114	–		–		–		–		–		–		–		–	
Steel, AISI 2315	1	1200	–		–		–		–		–		–		–		–	
Steel, AISI 2515	1	1198 1199	–		–		–		–		–		–		–		–	
Steel, AISI 3140	–		–		–		–		–		–		10	361	–		–	
Steel, AISI 4130	1	1153	–		–		–		–		–		–		–		–	
Steel, AISI 4140	1	1155	–		–		–		–		–		–		–		–	
Steel, AISI 4340	1	1213 1214	–		–		–		–		–		–		–		12	1177
Steel, AISI H 11	–		–		–		–		–		–		–		–		12	1146
Steel, Allegheny alloy No. 66	–		–		7	1180	–		–		–		–		–		–	

Substance Name	Thermal Conductivity		Specif. Heat		Thermal Radiative Properties								Thermal Diffusivity		Viscosity		Thermal Expansion	
					Emissivity		Reflectivity		Absorptivity		Transmissiv.							
	V.	Page	V.	Page	V.	Page	V.	Page	V.	Page	V.	Page	V.	Page	V.	Page	V.	Page
Steel, Allegheny metal		–		–	7	1225		–		–		–		–		–		–
Steel, aluminum	1	1142	4	626		–		–		–		–		–		–	12	639
Steel, AMS 2713	1	1210		–		–		–		–		–		–		–		–
Steel, AMS 2714	1	1213		–		–		–		–		–		–		–		–
Steel, AMS 6487		–		–		–		–		–		–		–		–	12	1146
Steel, antimony		–	4	629		–		–		–		–		–		–		–
Steel, Armco 21-6-9		–		–		–		–		–		–		–		–	12	1148
Steel, austenite		–	4	655		–		–		–		–		–		–		–
Steel, British	1	1187		–		–		–		–		–		–		–		–
Steel, British 4	1	1114		–		–		–		–		–		–		–		–
Steel, British 5	1	1114		–		–		–		–		–		–		–		–
Steel, British 7	1	1118		–		–		–		–		–		–		–		–
Steel, British, En 8	1	1184 1186		–		–		–		–		–		–		–		–
Steel, British, En 19	1	1153		–		–		–		–		–		–		–		–
Steel, British, En 31	1	1153 1154		–		–		–		–		–		–		–		–
Steel, British, En 32 A (BGK1)	1	1192		–		–		–		–		–		–		–		–
Steel, British, H.20	1	1154		–		–		–		–		–		–		–		–
Steel, British, H.27	1	1154		–		–		–		–		–		–		–		–
Steel, British, H.46	1	1154		–		–		–		–		–		–		–		–
Steel, British, Nicrosilal	1	1204		–		–		–		–		–		–		–		–
Steel, British, Staybrite	1	1161		–		–		–		–		–		–		–		–
Steel, carbon	1	1119 1126 1180 1185	4	619 623	7	935		–		–		–		–	10	332	11	573	12	1131 1133
Steel, carbon 2	1	1118		–		–		–		–		–		–		–		–
Steel, carbon 3	1	1118		–		–		–		–		–		–		–		–
Steel, carbon 4	1	1118		–		–		–		–		–		–		–		–
Steel, carbon, eutectoid		–	4	624		–		–		–		–		–		–		–
Steel, carbon, hyper eutectoid		–	4	624		–		–		–		–		–		–		–
Steel, carbon, U-8		–	4	624		–		–		–		–		–		–		–
Steel, carbon, Japanese	1	1185		–		–		–		–		–		–		–		–
Steel, carbon, SAE 1020		–		–		–		–		–		–		–		–	12	1167
Steel, Chromel 502	1	1210		–		–		–		–		–		–		–		–
Steel, chromium	1	1148 1152 1160 1164	4	632 638 678 687	7	1178 1190 1210 1231 1242 1253	7	938 1190 1196 1264 1283	7	1203 1206		–	10	338 344		–	12	707 1138
Steel, chromium, oxidized		–		–	9	1303		–		–		–		–		–		–
Steel, cobalt	1	1176	4	641		–		–		–		–		–		–	12	736 1158
Steel, cobalt, eutectoid		–	4	641		–		–		–		–		–		–		–
Steel, copper	1	1179	4	644		–		–		–		–		–		–	12	771 1162

| Substance Name | Thermal Conductivity | | Specif. Heat | | Thermal Radiative Properties | | | | | | | | | | | | | Thermal Diffusivity | | Viscosity | | Thermal Expansion | |
| | | | | | Emissivity | | Reflectivity | | Absorptivity | | Transmissiv. | | | | | | | | | | | |
	V.	Page	V.	Page	V.	Page	V.	Page	V.	Page	V.	Page	V.	Page	V.	Page	V.	Page
Steel, crucible	1	1204	–		–		–		–		–		–		–		–	
Steel, Cubex	–		–		–		–		–		–		–		–		12	870
Steel, eutectoid	–		4	655	–		–		–		–		–		–		–	
Steel, Fernichrome	–		–		–		–		–		–		–		–		12	1177 1178
Steel, Fernico	–		–		–		–		–		–		–		–		12	1178
Steel, fish-plate	1	1119	–		–		–		–		–		–		–		–	
Steel, FNCT	1	1213	–		–		–		–		–		–		–		–	
Steel, German, Krupp	1	1115 1184	–		–		–		–		–		–		–		–	
Steel, German PD4	1	1118	–		–		–		–		–		–		–		–	
Steel, German, St42.11	1	1186 1218	–		–		–		–		–		–		–		–	
Steel, GX 4881	–		–		–		–		–		–		10	342	–		–	
Steel, Haynes alloy N-155	1	1177	–		7	1227 1238	–		–		–		–		–		–	
Steel, high speed	1	1230 1231 1232 1234	–		–		–		–		–		–		–		–	
Steel, high speed 18	1	1233	–		–		–		–		–		–		–		–	
Steel, high speed 18-4-1	1	1233	–		–		–		–		–		–		–		–	
Steel, high speed M1	1	1195	–		–		–		–		–		–		–		–	
Steel, high speed M2	1	1233	–		–		–		–		–		–		–		–	
Steel, high speed M10	1	1195	–		–		–		–		–		–		–		–	
Steel, high speed T1	1	1233	–		–		–		–		–		–		–		–	
Steel, high-perm-49	1	1199	–		–		–		–		–		–		–		–	
Steel, HX 4249	–		–		–		–		–		–		10	342	–		–	
Steel, Incoloy	–		4	726	–		–		–		–		–		–		–	
Steel, Invar	1	1199	–		–		–		–		–		–		–		12	852 853 1175 1178 1179 1180 1181 1182
Steel, Invar 36	–		–		–		–		–		–		–		–		12	1183
Steel, Invar free cut	1	1205	–		–		–		–		–		–		–		–	
Steel, Japanese	1	1210	–		–		–		–		–		–		–		–	
Steel, Kanthal	–		–		7	1192	–		7	1204	–		–		–		–	
Steel, Kanthal, A	–		–		7	1192	–		–		–		–		–		–	
Steel, Kanthal, Oxidized	–		–		9	1303	–		–		–		–		–		–	
Steel, Kovar	1	1203	–		–		7	1313	–		–		–		–		–	
Steel, low alloy	1	1213	–		–		–		–		–		–		–		–	
Steel, low-exp-42	1	1205	–		–		–		–		–		–		–		–	
Steel, low Mn	1	1183	–		–		–		–		–		–		–		–	
Steel, M1 high speed tool	1	1195	–		–		–		–		–		–		–		–	
Steel, M10 high speed tool	1	1195	–		–		–		–		–		–		–		–	

Substance Name	Thermal Conductivity V.	Page	Specif. Heat V.	Page	Thermal Radiative Properties								Thermal Diffusivity V.	Page	Viscosity V.	Page	Thermal Expansion V.	Page
					Emissivity V.	Page	Reflectivity V.	Page	Absorptivity V.	Page	Transmissiv. V.	Page						
Steel, magnet, K.S.	1	1177	–		–		–		–		–		–		–		–	
Steel, manganese	1	1182 1191	4	647 650 655 723	7	1305 1307	–		7	1309	–		10	353 357	–		12	841 1165
Steel, manganese, eutectoid	–		4	655	–		–		–		–		–		–		–	
Steel, Maraging	–		–		–		–		–		–		–		–		12	1183
Steel, mild	2	1141	4	647	7	1305 1307	–		7	1310	–		–		–		–	
Steel, molybdenum	1	1194	–		–		–		–		–		–		–		12	845
Steel, nickel	1	1197 1202 1209 1212	4	660 665 726 729	7	942 1175 1317 1320	7	1312 1324	–		–		10	360 362	–		12	848 1175
Steel, nickel, oxidized	–		–		9	1305 1308	–		–		–		–		–		–	
Steel, Ni–Cr	1	1167 1168 1210 1213	–		–		–		–		–		–		–		–	
Steel, Nilo 36	–		–		–		–		–		–		–		–		12	1187
Steel, Nilo 40	–		–		–		–		–		–		–		–		12	1187
Steel, Niromet 42	–		–		–		–		–		–		–		–		12	853
Steel, Ni span	–		–		–		–		–		–		–		–		12	1183
Steel, Ni span C	1	1214	–		–		–		–		–		–		–		12	1183
Steel, oil-hardening non-deforming	1	1125	–		–		–		–		–		–		–		–	
Steel, oxidized	–		–		9	1305 1308	–		–		–		–		–		–	
Steel, pearlite	–		4	655	–		–		–		–		–		–		–	
Steel, phosphorus	1	1216	–		–		–		–		–		–		–		–	
Steel, platinum	–		–		–		–		–		–		–		–		12	861
Steel, Potomac A	–		–		7	1192	7	1198	–		–		–		–		–	
Steel, Russian	1	1118	–		–		–		–		–		–		–		–	
Steel, Russian, 15	–		–		–		–		–		–		10	354	–		–	
Steel, Russian, 22	1	1192 1218 1222	–		–		–		–		–		–		–		–	
Steel, Russian, 35	–		–		–		–		–		–		10	354	–		–	
Steel, Russian, 45	–		–		–		–		–		–		10	355	–		–	
Steel, Russian, EI-257	1	1166 1214	4	720	–		–		–		–		–		–		–	
Steel, Russian, EI-435	–		–		–		–		–		–		10	300	–		–	
Steel, Russian, EI-572	1	1167	–		–		–		–		–		–		–		–	
Steel, Russian, EI-606	1	1167	–		–		–		–		–		–		–		–	
Steel, Russian, ferrosilicon 45 percent	1	1218	–		–		–		–		–		–		–		–	
Steel, Russian, ferrotitanium	1	1225	–		–		–		–		–		–		–		–	
Steel, Russian, Kh Zn	1	1210	–		–		–		–		–		–		–		–	
Steel, Russian, R7	1	1236	–		–		–		–		–		–		–		–	
Steel, Russian, R10	1	1236	–		–		–		–		–		–		–		–	
Steel, Russian, R12	1	1236	–		–		–		–		–		–		–		–	

Substance Name	Thermal Conductivity		Specif. Heat		Thermal Radiative Properties								Thermal Diffusivity		Viscosity		Thermal Expansion	
					Emissivity		Reflectivity		Absorptivity		Transmissiv.							
	V.	Page	V.	Page	V.	Page	V.	Page	V.	Page	V.	Page	V.	Page	V.	Page	V.	Page
Steel, Russian, R15	1	1235	–		–		–		–		–		–		–		–	
Steel, Russian, R15Kh3	1	1235	–		–		–		–		–		–		–		–	
Steel, Russian, R15Kh3K5	1	1235	–		–		–		–		–		–		–		–	
Steel, Russian, R15Kh3K10	1	1235	–		–		–		–		–		–		–		–	
Steel, Russian, R15Kh3K12	1	1235 1236	–		–		–		–		–		–		–		–	
Steel, Russian, R15Kh4	1	1236	–		–		–		–		–		–		–		–	
Steel, Russian, R18	1	1236	–		–		–		–		–		–		–		–	
Steel, SAE 1010	1	1183	4	647	–		–		–		–		–		–		12	842 1165 1167
Steel, SAE 1015	1	1186	–		–		–		–		–		–		–		–	
Steel, SAE 1020	1	1183	–		–		–		–		–		10	354	–		12	1167
Steel, SAE 1095	1	1114	–		–		–		–		–		–		–		–	
Steel, SAE 4130	1	1153	–		–		–		–		–		10	339	–		–	
Steel, SAE 4140	1	1155	–		–		–		–		–		–		–		–	
Steel, SAE 4340	1	1213 1214	–		–		–		–		–		10	363 364	–		–	
Steel, silicon	1	1217	4	668 732	–		–		–		–		10	366	–		12	868
Steel, silver	1	1114	–		–		–		–		–		–		–		–	
Steel, soft	1	1126	–		–		–		–		–		–		–		–	
Steel, stainless	1	1148 1152 1160 1164	4	632 635 638 678 690 699 717	7	1178 1184 1190 1210 1231 1242 1256	7	1196 1264 1283	7	1203 1206	–		10	338 344	–		12	1138
Steel, stainless, 15-5PH	–		–		–		–		–		–		–		–		12	1141
Steel, stainless, 17-4PH	1	1168	4	717	–		–		–		–		–		–		12	1138 1141
Steel, stainless, 17-7	1	1165	–		–		–		–		–		–		–		–	
Steel, stainless, 17-7PH	1	1166	4	696	7	1223 1237 1245	7	1266 1268	7	1302	–		–		–		12	1138 1141
Steel, stainless, 18-8	1	1161 1162 1167 1168	–		7	1212 1214 1225 1226	–		7	1302	–		–		–		–	
Steel, stainless, 20Cr-25Ni	–		–		–		–		–		–		–		–		12	1175 1186
Steel, stainless, 347, oxidized	–		–		9	1305 1308	–		–		–		–		–		–	
Steel, stainless, 416	1	1168	–		–		–		–		–		–		–		–	
Steel, stainless, 3754	1	1161	–		–		–		–		–		–		–		–	
Steel, stainless, A-286	–		–		7	1322	7	1325	–		–		–		–		12	1175 1177
Steel, stainless, AFC-77	–		–		–		–		–		–		–		–		12	1145
Steel, stainless, AISI 202	–		–		–		–		–		–		10	339 340	–		–	
Steel, stainless, AISI 301	1	1165	4	693	7	1221 1226	7	1269 1288	7	1300	–		10	345 348	–		12	1138 1141 1142

Substance Name	Thermal Conductivity		Specif. Heat		Emissivity		Reflectivity		Absorptivity		Transmissiv.		Thermal Diffusivity		Viscosity		Thermal Expansion	
	V.	Page	V.	Page	V.	Page	V.	Page	V.	Page	V.	Page	V.	Page	V.	Page	V.	Page
Steel, stainless, AISI 301, corrugated sheets	-		-		-		-		-		-		10	552	-		-	
Steel, stainless, AISI 302	1	1161	-		7	1212 1213	-		7	1291	-		10	345	-		12	1138 1142
Steel, stainless, AISI 303	1	1165 1168	-		7	1212 1226 1254 1258 1259 1260	-		7	1297	-		-		-		12	1138 1142
Steel, stainless, AISI 304	1	1161 1165 1168	4	699	7	1213 1227 1244	7	1270			-		-		-		-	
Steel, stainless, AISI 304ELC	-		-		7	1213	-		-		-		-		-		-	
Steel, stainless, AISI 304L	-		-		-		-		-		-		-		-		12	1138 1142
Steel, stainless, AISI 305	-		4	702	-		-		-		-		-		-		-	
Steel, stainless, AISI 309	-		-		-		-		-		-		10	346	-		-	
Steel, stainless, AISI 310	1	1167 1168	4	705	7	1212 1213	-		-		-		-		-		12	1138 1142
Steel, stainless, AISI 316	1	1166	4	708	7	1221 1224 1237 1244	7	1266 1270 1271 1288	7	1300 130	-		10	347 348	-		12	1138 1143
Steel, stainless, AISI 321	-		-		7	1224 1237 1238 1244 1246	7	1266 1270 1272 1285	7	129	-		10	347	-		12	1138 1143
Steel, stainless, AISI 330	-		-		-		-		-		-		-		-		12	1177
Steel, stainless, AISI 347	1	1166 1168	4	711	7	1212 1222	7	1288	-		-		10	348	-		12	1138 1143 1144
Steel, stainless, AISI 403	1	1149	-		-		-		-		-		-		-		-	
Steel, stainless, AISI 406	-		-		-		-		-		-		-		-		12	1138 1144
Steel, stainless, AISI 410	1	1150	-		-		-		-		-		10	340	-		12	1138 1144
Steel, stainless, AISI 416	-		-		-		-		-		-		10	340 341	-		12	1138 1144
Steel, stainless, AISI 420	1	1162	4	678	-		-		-		-		-		-		12	1138 1144
Steel, stainless, AISI 422	-		-		-		-		-		-		-		-		12	1138 1145
Steel, stainless, AISI 430	1	1154	4	681	7	1193	-		-		-		10	341	-		-	
Steel, stainless, AISI 430F	-		-		-		-		-		-		-		-		12	1138 1145
Steel, stainless, AISI 440C	1	1154	-		-		-		-		-		-		-		12	1138 1145
Steel, stainless, AISI 446	1	1155 1156	4	684	7	1180 1187	7	1198	7	1207	-		10	341 342	-		12	1138 1145
Steel, stainless, AISI 455	-		-		-		-		-		-		-		-		12	1145
Steel, stainless, AISI 633	-		-		-		-		-		-		-		-		12	1145
Steel, stainless, AM 35	-		-		-		-		-		-		-		-		12	1145
Steel, stainless, AM 335	-		-		-		-		-		-		-		-		12	1145
Steel, stainless, AM 350	-		-		7	1225 1238 1245	7	1266 1267 1268	7	1302 1303	-		-		-		-	

Substance Name	Thermal Conductivity		Specif. Heat		Thermal Radiative Properties								Thermal Diffusivity		Viscosity		Thermal Expansion	
					Emissivity		Reflectivity		Absorptivity		Transmissiv.							
	V.	Page	V.	Page	V.	Page	V.	Page	V.	Page	V.	Page	V.	Page	V.	Page	V.	Page
Steel, stainless, AM 355	1	1168	4	717		–		–		–		–		–		–		–
Steel, stainless, AM 362		–		–		–		–		–		–		–		–	12	1145
Steel, stainless, AM 363		–		–		–		–		–		–		–		–	12	1145
Steel, stainless, AS21	1	1161		–		–		–		–		–		–		–		–
Steel, stainless, Ascoloy		–		–		–		–		–		–		–		–	12	1148
Steel, stainless, austenite		–	4	655		–		–		–		–		–		–	12	1138 1147
Steel, stainless, austenitic	1	1165 1183		–		–		–		–		–		–		–		–
Steel, stainless, British, Era ATV	1	1213		–		–		–		–		–		–		–		–
Steel, stainless, British, F.H.	1	1161		–		–		–		–		–		–		–		–
Steel, stainless, British, G 18B	1	1165 1213		–		–		–		–		–		–		–		–
Steel, stainless, British, Jessop G 17	1	1213		–		–		–		–		–		–		–		–
Steel, stainless, British, R20	1	1165		–		–		–		–		–		–		–		–
Steel, stainless, British, SF 11	1	1149		–		–	7	938		–		–		–		–		–
Steel, stainless, Carpenter 20-CB		–		–		–		–		–		–		–		–	12	1177
Steel, stainless, Crucible HMN		–	4	714		–		–		–		–		–		–		–
Steel, stainless, Crucible HNM	1	1168	4	~~714~~		–		–		–		–		–		–		–
Steel, stainless, French, Nimonic DS	1	1213		–		–		–		–		–		–		–		–
Steel, stainless, German X8CrNiMoNb 16 16		–		–		–		–		–		–	10	350 364		–		–
Steel, stainless, H-11, AMS 6487		–		–		–		–		–		–		–		–	12	1146
Steel, stainless, high alloy	1	1214		–		–		–		–		–		–		–		–
Steel, stainless, Kromare 55		–		–		–		–		–		–		–		–	12	1183
Steel, stainless, Macloy G	1	1213		–		–		–		–		–		–		–		–
Steel, stainless, Mark 1X 18 N9T		–	4	699		–		–		–		–		–		–		–
Steel, stainless, Mark 12 MX		–	4	723		–		–		–		–		–		–		–
Steel, stainless, N-155		–		–	7	1180 1181 1186 1222 1226 1236	7	1198 1266	7	1207 130		–		–		–		–
Steel, stainless, Nimonic PE7	1	1206		–		–		–		–		–		–		–		–
Steel, stainless, oxidized		–		–	9	1303		–		–		–		–		–		–
Steel, stainless, PH14-8Mo		–		–		–		–		–		–		–		–	12	1146
Steel, stainless, PH15-7Mo		–		–	7	1223 1237 1245	7	1266 1267	7	1301		–		–		–		–
Steel, stainless, Rex 78	1	1213		–		–		–		–		–		–		–		–
Steel, stainless, Russian	1	1150 1161		–		–		–		–		–		–		–		–
Steel, stainless, Russian, 0Kh 16N 36V 3T		–	4	726		–		–		–		–		–		–		–
Steel, stainless, Russian, 0Kh 20N 60B		–		–		–		–		–		–	10	299		–		–
Steel, stainless, Russian, 0Kh 21N 78T		–		–		–		–		–		–	10	300		–		–
Steel, stainless, Russian, 1Kh 14N 14V 2M	1	1166 1214		–		–		–		–		–		–		–		–
Steel, stainless, Russian, 1Kh 18N 9T	1	1168	4	699		–		–		–		–		–		–		–

Substance Name	Thermal Conductivity		Specif. Heat		Thermal Radiative Properties								Thermal Diffusivity		Viscosity		Thermal Expansion	
					Emissivity		Reflectivity		Absorptivity		Transmissiv.							
	V.	Page	V.	Page	V.	Page	V.	Page	V.	Page	V.	Page	V.	Page	V.	Page	V.	Page
Steel, stainless, Russian, 4Kh 13		–	4	690		–		–		–		–		–		–		–
Steel, stainless, Russian, 5 ZA 2	1	1213		–		–		–		–		–		–		–		–
Steel, stainless, Russian, 12MKH	1	1192		–		–		–		–		–		–		–		–
Steel, stainless, Russian, 15Kh 12VMF	1	1156		–		–		–		–		–		–		–		–
Steel, stainless, Russian, EI-802	1	1156 1157		–		–		–		–		–		–		–		–
Steel, stainless, Russian, EI-855	1	1214	4	726		–		–		–		–		–		–		–
Steel, stainless, Russian, EYA 1T	1	1168		–		–		–		–		–		–		–		–
Steel, stainless, Russian, EYA 2	1	1166		–		–		–		–		–		–		–		–
Steel, stainless, Russian, WF 100	1	1166		–		–		–		–		–		–		–		–
Steel, stainless, SF 20		–		–		–	7	1266		–		–		–		–		–
Steel, T-261		–	4	655		–		–		–		–		–		–		–
Steel, T-262		–	4	655		–		–		–		–		–		–		–
Steel, T-270		–	4	655		–		–		–		–		–		–		–
Steel, T-278		–	4	655		–		–		–		–		–		–		–
Steel, T-279		–	4	655		–		–		–		–		–		–		–
Steel, T-310		–	4	655		–		–		–		–		–		–		–
Steel, T-311		–	4	655		–		–		–		–		–		–		–
Steel, tin		–	4	672		–		–		–		–		–		–		–
Steel, titanium	1	1225	4	675 735		–		–		–		–		–		–		–
Steel, tool	1	1115 1233		–		–		–		–		–		–		–		–
Steel, tool 1.1 C		–		–		–		–		–		–	10	335		–		–
Steel, tungsten	1	1226	4	738	7	945		–		–		–		–		–	12	1199
Steel, vascojet 1000		–		–	7	1192	7	1199		–		–		–		–		–
Steatite, 10 B2	2	853		–		–		–		–		–		–		–		–
Stellite		–		–		–	7	1154		–		–		–		–		–
Stellite 3		–		–		–		–		–		–		–		–	12	1067 1069
Stellite 6		–		–		–		–		–		–		–		–	12	1067 1069
Stellite 21		–		–		–		–		–		–		–		–	12	1067 1069
Stellite 23		–		–		–		–		–		–		–		–	12	1067 1069
Stellite 25		–		–		–		–		–		–		–		–	12	1067 1069
Stellite 27		–		–		–		–		–		–		–		–	12	1067 1069
Stellite 30		–		–		–		–		–		–		–		–	12	1067 1070
Stellite 31		–		–		–		–		–		–		–		–	12	1067 1070
Stellite HE 1049		–	4	526		–		–		–		–		–		–		–
Stibium	1	10	4	6		–		–		–		–		–		–		–
Stilbite		–		–		–		–		–		–	8	1694		–		–
Strawberry		–		–		–		–		–		–		–	10	645		–

Substance Name	Thermal Conductivity		Specif. Heat		Thermal Radiative Properties								Thermal Diffusivity		Viscosity		Thermal Expansion	
					Emissivity		Reflectivity		Absorptivity		Transmissiv.							
	V.	Page	V.	Page	V.	Page	V.	Page	V.	Page	V.	Page	V.	Page	V.	Page	V.	Page
Strontia	2	194	5	225	–		–		–		–		–		–		–	
Strontium, Sr	–		4	218	–		–		–		–		10	170	–		12	313
Strontium aluminum silicate, SrAl$_2$Si$_2$O$_8$	–		–		–		–		–		–		–		–		13	728
Strontium boride, SrB$_6$	–		–		8	732	–		–		–		–		–		–	
Strontium bromide, SrBr$_2$	–		5	775	–		–		–		–		–		–		–	
Strontium carbonate, SrCO$_3$	–		5	1136	–		–		–		–		–		–		–	
Strontium chloride, SrCl$_2$	–		5	890	–		–		–		–		–		–		13	1014
Strontium fluoride, SrF$_2$	–		5	1003	–		8	968	–		8	971	–		–		13	1065
Strontium fluoride + ΣXi, mixture	2	791	–		–		–		–		–		–		–		–	
Strontium hafnium oxide, SrO·HfO$_2$	–		–		–		8	597	–		–		–		–		13	502
Strontium lead oxide, SrO·PbO$_2$	–		–		–		–		–		–		–		–		13	516
Strontium metatitanate + cobalt, cermet	2	722	–		–		–		–		–		–		–		–	
Strontium molybdenum oxide, SrO·MoO$_3$	–		–		–		–		–		–		–		–		13	522
Strontium nitrate, Sr(NO$_3$)$_2$	–		5	1154	–		–		–		–		–		–		13	668
Strontium oxide, SrO	2	194	5	225	–		8	546	–		–		–		–		13	372
Strontium oxide + titanium oxide + ΣXi, mixture	2	517	–		–		–		–		–		–		–		–	
Strontium oxide + zinc oxide + ΣXi, mixture	2	520	–		–		–		–		–		–		–		–	
Strontium silicates:																		
SrSiO$_3$	–		5	1605	–		–		–		–		–		–		–	
Sr$_2$SiO$_4$	–		5	1608	–		–		–		–		–		–		–	
Strontium silicide, Sr$_2$Si	1	1343	–		–		–		–		–		–		–		–	
Strontium stannide, Sr$_2$Sn	1	1344	–		–		–		–		–		–		–		–	
Strontium sulfide, SrS	–		5	708	–		–		–		–		–		–		13	1240
Strontium-tin intermetallic compound, Sr$_2$Sn	1	1344	–		–		–		–		–		–		–		–	
Strontium titanium oxides:																		
SrO·TiO$_2$	2	304	5	1611	8	651	8	653	–		8	657	–		–		13	570
2SrO·TiO$_2$	–		5	1614	–		–		–		–		–		–		–	
Strontium zirconium oxide, SrO·ZrO$_2$	2	307	5	1617	–		8	676	–		–		–		–		13	611
Stycast 1266	–		–		–		–		–		–		–		–		13	1505
Stycast 2850	–		–		–		–		–		–		–		–		13	1505
Stycast 2850 FT	–		–		–		–		–		–		–		–		13	1504
Stycast 2850 GT	–		–		–		–		–		–		–		–		13	1504
Styrene	–		6s	84	–		–		–		–		–		–		–	
Styrofoam polystyrene	2	965	–		–		–		–		–		–		–		–	
Sucrose, solution	–		–		–		–		–		–		10	591	–		–	
Sulfides, miscellaneous	–		–		8	1231	8	1233	–		8	1235	–		–		–	
Sulfothiorine, Na$_2$S$_2$O$_3$·5H$_2$O	2	693	–		–		–		–		–		–		–		–	
Sulfur, S	2	89	5 / 6s	21 / 85	–		8	115	–		8	121	10	171	–		13	162
Sulfur, S$_2$	–		6s	85	–		–		–		–		–		–		–	
Sulfur dichloride	–		6s	87	–		–		–		–		–		–		–	

Substance Name	Thermal Conductivity		Specif. Heat		Thermal Radiative Properties								Thermal Diffusivity		Viscosity		Thermal Expansion	
					Emissivity		Reflectivity		Absorptivity		Transmissiv.							
	V.	Page	V.	Page	V.	Page	V.	Page	V.	Page	V.	Page	V.	Page	V.	Page	V.	Page
Sulfur difluoride	–		6s	87	–		–		–		–		–		–		–	
Sulfur dioxide, SO_2	3	116	6	97	–		–		–		–		–		11	91	–	
Sulfur dioxide + sulfuryl fluoride, mixture	–		–		–		–		–		–		–		11	570	–	
Sulfur, flowers of sulfur	–		–		–		8	117	–		–		–		–		–	
Sulfur hexafluoride	–		6s	87	–		–		–		–		–		–		–	
Sulfur monochloride	–		6s	88	–		–		–		–		–		–		–	
Sulfur monoxide, SO	–		6s	88	–		–		–		–		–		–		–	
Sulfur oxychloride	–		6s	90	–		–		–		–		–		–		–	
Sulfur tetrafluoride	–		6s	88	–		–		–		–		–		–		–	
Sulfur trioxide, SO_3	–		6s	88	–		–		–		–		–		–		–	
Sulfuric anhydride	–		6s	88	–		–		–		–		–		–		–	
Sulfuric ether	–		6	194	–		–		–		–		–		–		–	
Sulfurous acid anhydride, SO_2	3	116	–		–		–		–		–		–		–		–	
Sulfurous oxychloride	–		6s	90	–		–		–		–		–		–		–	
Sulfuryl fluoride	–		6s	89	–		–		–		–		–		–		–	
Superinvar	–		–		–		–		–		–		–		–		12	1182 1183
Superlith XXXN	–		–		–		8	1219	8	1226	–		–		–		–	
Superpax	–		–		–		8	613	–		–		–		–		–	
Sylvan	–		6s	63	–		–		–		–		–		–		–	
Systems, honeycomb structures, metallic-nonmetallic	2	1015	–		–		–		–		–		–		–		–	
Systems, honeycomb structures, nonmetallic	2	1010	–		–		–		–		–		–		–		–	
Talc	–		–		–		–		–		8	1694	–		–		–	
Tantalum, Ta	1	355	4	221	7	654 661 666 672	7	678 684	7	687	–		10	173	–		12	316
Tantalum, anodized	–		–		9	1284	9	1286	9	1287	–		–		–		–	
Tantalum alloys:																		
Ta + Nb	1	801	–		–		–		–		–		10	256	–		12	950
Ta + W	1	802	4	434	7	1021 1024	–		–		–		10	258	–		12	979
Ta + Nb + ΣXi	1	1062	4	592	–		–		–		–		10	320	–		12	1266 1268 1269
Ta + W + ΣXi	1	1065	4	595	7	1481	–		–		–		10	322	–		12	1267 1268 1270
Ta + W + ΣXi, T-111	–		–		7	1481	–		–		–		10	323	–		12	1267 1270
Ta + W + ΣXi, T-222	1	1066	–		–		–		–		–		10	323	–		12	1270
Tantalum aluminum compound + tantalum, mixture	–		–		–		8	1433	–		–		–		–		–	
Tantalum-aluminum intermetallic compound, $TaAl_3$	–		–		8	1330	8	1332	–		–		–		–		–	

Substance Name	Thermal Conduc-tivity		Specif. Heat		Thermal Radiative Properties								Thermal Diffu-sivity		Visco-sity		Thermal Expan-sion	
					Emis-sivity		Reflec-tivity		Absorp-tivity		Trans-missiv.							
	V.	Page	V.	Page	V.	Page	V.	Page	V.	Page	V.	Page	V.	Page	V.	Page	V.	Page
Tantalum beryllides:																		
TaBe₁₂		–	5	322		–		–		–		–		–		–		–
Ta₂Be₁₇		–	5	325		–		–		–		–		–		–		-
Tantalum borides:																		
TaB		–	5	372		–		–		–		–		–		–	13	797
TaB₂	1	1345	5	368		–		–		–		–		–		–	13	772
Tantalum carbides:																		
TaC	2	589	5	451	8	811 813 815 817		–		–		–	10	483		–	13	879
Ta₂C		–		–		–		–		–		–		–		–	13	936
Tantalum carbide + tungsten carbide, mixture		–		–		–		–		–		–		–		–	13	940
Tantalum-germanium intermetallic compound, TaGe₂	1	1348		–		–		–		–		–		–		–		–
Tantalum nitrides:																		
TaN	2	665	5	1090	8	1072 1074 1076		–		–		–		–		–	13	1162
Ta₂N		–		–	8	1075		–		–		–		–		–		–
Tantalum oxide, Ta₂O₅		–	5	228	8	427 428	8	430		–		–		–		–	13	374
Tantalum oxide + beryllium tantalum compound, cermet		–		–	8	1403 1404	8	1406		–		–		–		–		–
Tantalum phosphate, TaPO₅		–		–		–		–		–		–		–		–	13	690
Tantalum silicide, TaSi₂		–	5	598	8	1156 1157 1159	8	1161		–		–		–		–	13	1202
Tantalum vanadium oxide, Ta₂O₅·V₂O₅		–		–		–		–		–		–		–		–	13	597
Tar camphor		–	6s	69		–		–		–		–		–		–		–
Teak	2	1087		–		–		–		–		–	10	646		–		–
Teak, burmese		–		–		–		–		–		–	10	646		–		–
Technetium	1	363		–		–		–		–		–	10	178		–		–
Teflon	2	967		–		–		–		–		–	10	609		–	13	1445
Teflon 5		–		–		–		–		–		–		–		–	13	1443
Teflon 6		–		–		–		–		–		–		–		–	13	1445
Teflon, AMS 3651		–		–		–	8	1733		–		–		–		–		–
Teflon, BMS-a-71		–		–		–	8	1733		–		–		–		–		–
Teflon, Duroid 5600	2	968		–		–		–		–		–		–		–		–
Teflon I		–		–		–		–		–		–		–		–	13	1445
Teflon film		–		–		–		–		–		–		–		–	13	1447
Teflon TF1		–		–		–		–		–		–		–		–	13	1447
Tekite, synthetic		–		–		–		–		–		–	10	579		–		–
Tellurac-cured butyl	2	983		–		–		–		–		–		–		–		–
Telluric acid anhydride		–	5	231		–		–		–		–		–		–		–
Tellurite		–	5	231		–		–		–		–		–		–		–

Substance Name	Thermal Conductivity		Specif. Heat		Thermal Radiative Properties								Thermal Diffusivity		Viscosity		Thermal Expansion	
					Emissivity		Reflectivity		Absorptivity		Transmissiv.							
	V.	Page	V.	Page	V.	Page	V.	Page	V.	Page	V.	Page	V.	Page	V.	Page	V.	Page
Tellurium, Te	1	366	4	229		–	8	123 125 128		–	8	130 136	10	181		–	13	163
Tellurium alloys:																		
Te + Se	1	805		–		–		–		–		–		–		–		–
Te + Tl	1	808		–		–		–		–		–		–		–		–
Te + As + ΣXi	1	1068		–		–		–		–		–		–		–		–
Tellurium dioxide, TeO_2		–	5	231	8	432	8	434		–	8	436		–		–		–
Terbium, Tb	1	372	4	232		–		–		–		–	10	182		–	12	323
Terbium alloy, Tb + Y		–		–		–		–		–		–		–		–	12	983
Terbium borides:																		
TbB_6		–		–	8	723		–		–		–		–		–		–
TbB_{12}		–		–		–		–		–		–		–		–	13	791
Terbium carbide, TbC_2		–		–		–		–		–		–		–		–	13	936
Terylane filament yarn		–		–		–		–		–		–		–		–	13	1438
Tetrabromoactylene		–	6s	90		–		–		–		–		–		–		–
Tetrabromomethane		–	6s	15		–		–		–		–		–		–		–
sym-Tetrabromoethane		–	6s	90		–		–		–		–		–		–		–
1,1,2,2-Tetrabromoethane		–	6s	90		–		–		–		–		–		–		–
sym-Tetrachloroethane		–	6s	90		–		–		–		–		–		–		–
1,1,2,2-Tetrachloroethane		–	6s	90		–		–		–		–		–		–		–
1,1,2,2-Tetrachloro-1,2-difluoroethane		–	6s	90		–		–		–		–		–		–		–
Tetrachloroethylene		–	6 6s	159 90		–		–		–		–		–		–		–
Tetrachlorosilane		–	6s	83		–		–		–		–		–		–		–
Tetradecane		–	6s	90		–		–		–		–		–		–		–
Tetradeuteriomethane		–	6s	58		–		–		–		–		–		–		–
Tetrafluorosilane		–	6s	84		–		–		–		–		–		–		–
1,2,3,4-Tetrahydrobenzene		–	6s	25		–		–		–		–		–		–		–
1,2,3,4-Tetramethylbenzene		–	6s	90		–		–		–		–		–		–		–
1,2,3,4-tetramethylbenzene, prehnitene		–	6s	90		–		–		–		–		–		–		–
1,2,3,4-tetramethylbenzene, prehnitole		–	6s	90		–		–		–		–		–		–		–
1,2,3,4-tetramethylbenzene, prenitol		–	6s	90		–		–		–		–		–		–		–
1,2,3,5-Tetramethylbenzene		–	6s	90		–		–		–		–		–		–		–
1,2,4,5-Tetramethylbenzene		–	6s	90		–		–		–		–		–		–		–
1,2,4,5-Tetramethylbenzene, durene		–	6s	90		–		–		–		–		–		–		–
Tetramethylmethane		–	6s	33		–		–		–		–		–		–		–
Tetratritiomethane		–	6s	58		–		–		–		–		–		–		–
Tetrol		–	6s	43		–		–		–		–		–		–		–
Textolite plastic		–		–		–		–		–		–	10	559		–		–
TFE-fluorocarbon		–		–		–		–		–		–	10	588		–		–
Thallium, Tl	1	376	4	237		–	7	690 693		–	7	696	10	183		–	12	328

Substance Name	Thermal Conduc-tivity		Specif. Heat		Thermal Radiative Properties								Thermal Diffu-sivity		Visco-sity		Thermal Expan-sion	
					Emis-sivity		Reflec-tivity		Absorp-tivity		Trans-missiv.							
	V.	Page	V.	Page	V.	Page	V.	Page	V.	Page	V.	Page	V.	Page	V.	Page	V.	Page
Thallium alloys:																		
Tl + Cd	1	811	–		–		–		–		–		–		–		–	
Tl + In	1	812	–		–		–		–		–		–		–		12	832
Tl + Pb	1	815	4	437	–		–		–		–		–		–		–	
Tl + Sn	1	821	–		–		–		–		–		–		–		–	
Tl + Te	1	818	–		–		–		–		–		–		–		–	
Thallium bromide, TlBr	2	570	–		–		8	778 780 782	–		8	783	–		–		13	826
Thallium bromide chloride	–		–		–		8	1455	–		8	1457	–		–		–	
Thallium bromide iodide	–		–		8	1459	8	1461	–		8	1463	–		–		–	
Thallium bromide + thallium chloride, KRS-6	–		–		–		8	1455	–		8	1457	–		–		13	1015
Thallium bromide + thallium chloride, mixture	–		–		–		8	1455	–		8	1457	–		–		–	
Thallium bromide + thallium iodide, KRS-5	–		–		8	1459	8	1461	–		8	1463	–		–		–	
Thallium bromide + thallium iodide, mixture	–		–		8	1459	8	1461	–		8	1463	–		–		13	1123
Thallium carbide, TlC	2	625	–		–		–		–		–		–		–		–	
Thallium chloride, TlCl	–		–		–		8	899 901	–		8	903	–		–		13	1014
Thallium chloride + thallium bromide, mixture	–		–		–		8	1455	–		8	1457	–		–		–	
Thallium monofluoride, TlF	–		–		–		–		–		–		–		–		13	1069
Thallium monohydrogen difluoride, TlHF₃	–		5	1006	–		–		–		–		–		–		–	
Thallium iodide, TlI	–		–		–		–		–		8	1029	–		–		13	1121
Thallium-lead intermetallic compound, Tl₂Pb	1	1349	–		–		–		–		–		–		–		–	
Thallium nitrate, TlNO₃	–		5	1157	–		–		–		–		–		–		–	
2-Thiapropane	–		6s	69	–		–		–		–		–		–		–	
Thiocarbonyl chloride	–		6s	91	–		–		–		–		–		–		–	
Thionyl chloride	–		6s	90	–		–		–		–		–		–		–	
Thionyl fluoride	–		6s	91	–		–		–		–		–		–		–	
Thiophosgene	–		6s	91	–		–		–		–		–		–		–	
2,5-Thioxene	–		6s	33	–		–		–		–		–		–		–	
Thoria	2	195	5	234	8	438 440	8	442 444	8	446	–		–		–		–	
Thorium, billet A	–		–		–		–		–		–		–		–		12	334
Thorium, billet MX	–		–		–		–		–		–		–		–		12	334
Thorium, Th	1	381	4	242	7	699 701	–		–		–		10	184	–		12	332
Thorium alloys:																		
Th + Ce	–		–		–		–		–		–		–		–		12	702
Th + U	1	822	–		–		–		–		–		–		–		–	
Thorium borides:																		
ThB₄	–		5	375	8	730	–		–		–		–		–		13	797
ThB₆	–		–		–		–		–		–		–		–		13	797

Substance Name	Thermal Conductivity		Specif. Heat		Thermal Radiative Properties										Thermal Diffusivity		Viscosity		Thermal Expansion	
					Emissivity		Reflectivity		Absorptivity		Transmissiv.									
	V.	Page	V.	Page	V.	Page	V.	Page	V.	Page	V.	Page	V.	Page	V.	Page	V.	Page	V.	Page
Thorium carbides:																				
ThC	2	592		–	8	852		–		–		–		–				–	13	937
ThC$_2$	2	593		–	8	852		–		–		–		–				–	13	886
Nonstoichiometric		–	5	454		–		–		–		–		–				–		–
Thorium monocarbide + uranium monocarbide, ThC + UC, mixt		–		–		–		–		–		–		–				–	13	941
Thorium tetrafluoride, ThF$_4$		–	5	1009		–		–		–		–		–				–		–
Thorium nitride, ThN		–		–		–		–		–		–		–				–	13	1163
Thorium oxide, ThO$_2$	2	195	5	234	8	438 440	8	442 444	8	446		–	10	401				–	13	376
Thorium oxide + molybdenum, cermet	1	1429		–		–		–		–		–		–						–
Thorium oxide + nickel, cermet		–		–	8	1408		–		–		–		–						–
Thorium oxide + tungsten, cermet	1	1439		–		–		–		–		–		–						–
Thorium oxide + uranium oxide, mixture	2	413		–		–		–		–		–		–						–
Thorium phosphide, ThP		–		–		–		–		–		–		–				–	13	1183
Thorium silicate, ThSiO$_3$		–		–		–		–		–		–		–				–	13	728
Thorium sulfide, ThS		–		–		–		–		–		–		–				–	13	1240
Thorium disulfide, ThS$_2$		–	5	711		–		–		–		–		–						
Thoron	3	84		–		–		–		–		–		–				–		–
Thulia		–		–	8	447 450		–		–		–		–				–		–
Thulium, Tm	1	385	4	245		–		–		–		–	10	187			12	336
Thulium borides:																				
TmB$_6$		–		–	8	723		–		–		–		–				–		–
TmB$_{12}$		–		–		–		–		–		–		–				–	13	791
Thulium oxide, Tm$_2$O$_3$		–		–	8	447 450		–		–		–		–				–	13	383
Tin, Sn	1	389	4	249	7	703 705	7	707	7	710 712 714 717	7	720	10	188				–	12	339
Tin, gray		–	4	249		–		–		–		–		–				–	12 12	341 341
Tin, white		–	4	249		–		–		–		–		–				–		–
Tin anhydride, SnO$_2$	2	199		–		–		–		–		–		–				–		–
Tin alloys:																				
Sn + Ag	1	845		–		–		–		–		–		–				–		–
Sn + Al	1	823		–		–		–		–		–		–				–		–
Sn + Bi	1	827	4	440		–		–		–		–		–				–	12	684
Sn + Cd	1	830		–		–		–		–		–		–				–	12	697
Sn + Cu	1	833		–		–		–		–		–		–				–		–
Sn + In	1	834	4	443		–		–	7	1026		–		–				–	12	835
Sn + Hg	1	842		–		–		–		–		–		–				–		–
Sn + Mg		–	4	449		–		–		–		–		–				–	12	884
Sn + Pb	1	839	4	446		–		–		–		–		–				–	12	872

Substance Name	Thermal Conductivity		Specif. Heat		Thermal Radiative Properties								Thermal Diffusivity		Viscosity		Thermal Expansion	
					Emissivity		Reflectivity		Absorptivity		Transmissiv.							
	V.	Page	V.	Page	V.	Page	V.	Page	V.	Page	V.	Page	V.	Page	V.	Page	V.	Page
Tin alloys: (continued)																		
Sn + Sb	1	824	–		–		–		–		–		–		–		–	
Sn + Tl	1	846	–		–		–		–		–		–		–		–	
Sn + Zn	1	847	–		–		–		–		–		–		–		–	
Sn + Cu + ΣXi	1	1072	–		–		–		–		–		–		–		–	
Sn + Sb + ΣXi	1	1069	–		–		–		–		–		–		–		–	
Sn + Sb + ΣXi, bearing metal, white	1	1070	–		–		–		–		–		–		–		–	
Sn + Sb + ΣXi, SAE bearing alloy 11	1	1070	–		–		–		–		–		–		–		–	
Sn + Sb + ΣXi, SAE bearing alloy 10	1	1070	–		–		–		–		–		–		–		–	
Tin ash, SnO_2	2	199	–		–		–		–		–		–		–		–	
Tin + copper coating on steel substrate	–		–		–		9	757	–		–		–		–		–	
Tin tetrachloride, butter of tin	–		6s	91	–		–		–		–		–		–		–	
Tin tetrachloride, tin crystal	–		6s	91	–		–		–		–		–		–		–	
Tin dioxide + zinc oxide, mixture	2	419 438	–		–		–		–		–		–		–		–	
Tin dioxide + zinc oxide + ΣXi, mixture	2	524 528	–		–		–		–		–		–		–		–	
Tin oxides:																		
SnO	–		5	237	–		–		–		–		–		–		–	
SnO_2	2	199	5	240	–		8	451	–		–		–		–		13	386
SnO_2, flowers of tin	2	199	5	240	–		–		–		–		–		–		–	
Tin phosphate, $Sn_3(PO_4)_2$	–		–		–		–		–		–		–		–		13	691
Tin salt	–		6s	91	–		–		–		–		–		–		–	
Tin selenide, $SnSe_2$	1	1352	–		–		–		–		–		–		–		13	1192
Tin sulfide	–		–		–		8	1233	–		–		–		–		–	
Tin telluride, SnTe	1	1355	–		–		8	1259	–		8	1264	–		–		13	1260
Tin tetrachloride	–		6s	91	–		–		–		–		–		–		–	
Titania	2	202	5	246	8	456 458	8	461 473	8	475	8	476	–		–		–	
Titania, dense	2	204	–		–		–		–		–		–		–		–	
Titanic oxide, TiO_2	2	202	–		–		–		–		–		–		–		–	
Titanium, Ti	1	410	4	257	7	723 726 729 732 735 738	7	744 751 769	7	771	7	773	10	194	–		12	346
Titanium, 0.5 percent impurities	–		4	257	–		–		–		–		–		–		–	
Titanium, MSM-70	–		4	257	–		–		–		–		–		–		–	
Titanium, oxidized	–		–		–		9	1322	–		–		–		–		–	
Titanium, RS-70	–		4	257	–		–		–		–		–		–		–	
Titanium, Ti-75A	–		4	257	7	724 727	7	746	–		–		–		–		–	
Titanium, unknown impurities	1	1089	–		–		–		–		–		–		–		–	

Substance Name	Thermal Conduc-tivity		Specif. Heat		Thermal Radiative Properties								Thermal Diffu-sivity		Visco-sity		Thermal Expan-sion	
					Emis-sivity		Reflec-tivity		Absorp-tivity		Trans-missiv.							
	V.	Page	V.	Page	V.	Page	V.	Page	V.	Page	V.	Page	V.	Page	V.	Page	V.	Page
Titanium alloys:																		
Ti + Al	1	848		–		–		–		–		–	10	260		–	12	659
Ti + Al, A-110 '	1	1074		–		–		–		–		–		–		–		–
Ti + Al, BT-5		–		–		–		–		–		–	10	261		–		–
Ti + Al, Russian, 48-OT-3		–		–		–		–		–		–		–		–	13	752
Ti + Cr		–		–		–		–		–		–		–		–	12	726
Ti + Mo		–	4	456		–		–		–		–		–		–	12	910
Ti + Mo, M-6		–	4	456		–		–		–		–		–		–		–
Ti + Mo, M-8		–	4	456		–		–		–		–		–		–		–
Ti + Mo, M-9		–	4	456		–		–		–		–		–		–		–
Ti + Mo, M-10		–	4	456		–		–		–				–		–		–
Ti + Mn	1	849	4	453	7	1028 1032	7	1037	7	1041		–		–		–	12	894
Ti + Mn, AMS 4908		–		–	7	1030	7	1038		–		–		–		–		–
Ti + Mn, AMS 4908A	1	850		–		–		–		–		–		–		–		–
Ti + Mn, ASTM B265-58T, grade 3 and 4		–	4	257		–		–		–		–		–		–		–
Ti + Mn, ASTM B265-58T, grade 6	1	1074		–		–		–		–		–		–		–		–
Ti + Mn, ASTM B265-58T, grade 7	1	850		–		–		–		–		–		–		–		–
Ti + Mn, C-110M	1	850	4	453	7	1030 1034	7	1038	7	1041		–		–		–		–
Ti + Mn, MST-8Mn	1	850	4	453		–		–		–		–		–		–		–
Ti + Mn, MSM-8Mn		–	4	453		–		–		–		–		–		–		–
Ti + Mn, RS-110 A		–	4	453		–		–		–		–		–		–		–
Ti + Mn, RC-130A		–	4	453		–		–		–		–		–		–		–
Ti + Mn, Ti-130A	1	850		–		–		–		–		–		–		–		–
Ti + Mn, Ti-8Mn	1	850	4	453		–		–		–		–		–		–		–
Ti + O	1	852		–		–		–		–		–		–		–		–
Ti + V		–		–		–		–		–		–		–		–	12	990
Ti + W		–		–		–		–		–		–		–		–	12	988
Ti + Zr		–		–		–		–		–		–		–		–	12	993
Ti + Al + ΣXi	1	1073	4	598	7	1483 1486 1490	7	1497	7	1500		–	10	325		–	12	1272
Ti + Al + ΣXi, AMS 4925A	1	1074 1084		–		–		–		–		–		–		–		–
Ti + Al + ΣXi, AMS 4926	1	1074		–		–		–		–		–		–		–		–
Ti + Al + ΣXi, AMS 4928	1	1074	4	598		–		–		–		–		–		–		–
Ti + Al + ΣXi, AMS 4929	1	1074		–		–		–		–		–		–		–		–
Ti + Al + ΣXi, AMS 4969	1	1074		–		–		–		–		–		–		–		–
Ti + Al + ΣXi, anodized		–		–		–	9	1289		–		–		–		–		–
Ti + Al + ΣXi, anodized, A-110 AT		–		–		–	9	1289		–		–		–		–	12	1274 1280
Ti + Al + ΣXi, ASTM B265-58T, grade 6	1	1074		–		–		–		–		–		–		–		–
Ti + Al + ΣXi, C-120AV		–	4	598		–		–		–		–		–		–	12	1274
Ti + Al + ΣXi, C-130AM	1	1074		–		–		–		–		–		–		–		–

| Substance Name | Thermal Conductivity | | Specif. Heat | | Thermal Radiative Properties | | | | | | | | Thermal Diffusivity | | Viscosity | | Thermal Expansion | |
| | | | | | Emissivity | | Reflectivity | | Absorptivity | | Transmissiv. | | | | | | | |
	V.	Page	V.	Page	V.	Page	V.	Page	V.	Page	V.	Page	V.	Page	V.	Page	V.	Page
Titanium alloys: (continued)																		
Ti + Al + ΣXi, MSM-4Al-4Mn	1	1074 1084	–		–		–		–		–		–		–		–	
Ti + Al + ΣXi, MSM-6Al-4V	1	1074	4	598	–		–		–		–		–		–		–	
Ti + Al + ΣXi, MST-6Al-4V	1	1074	4	598	–		–		–		–		–		–		–	
Ti + Al + ΣXi, RC-130B	–		–		–		–		–		–		–		–		12	1274
Ti + Al + ΣXi, TA5E	–		–		7	1484	–		–		–		–		–		–	
Ti + Al + ΣXi, TA6V	–		–		7	1484	–		–		–		–		–		–	
Ti + Al + ΣXi, Ti-155A	1	1074	–		–		–		–		–		–		–		–	
Ti + Al + ΣXi, Ti-4Al-3Mo-1V	1	1074 1075	4	598	–		–		–		–		–		–		12	1274 1275
Ti + Al + ΣXi, Ti-4Al-4Mn	1	1074 1084	–		–		–		–		–		–		–		–	
Ti + Al + ΣXi, Ti-5Al-1.4Cr-1.5Fe-1.2Mo	1	1074	–		–		–		–		–		–		–		–	
Ti + Al + ΣXi, Ti-5Al-2.5Sn	1	1074	–		–		–		–		–		–		–		12	1274 1280
Ti + Al + ΣXi, Ti-5Al-5Sn-5Zr	–		–		–		–		–		–		–		–		12	1275 1276
Ti + Al + ΣXi, Ti-5Al-5Sn-5Zr-1Mo-1V	–		–		–		–		–		–		–		–		12	1276 1277
Ti + Al + ΣXi, Ti-6Al-2Sn-4Zr-2Mo	–		–		–		–		–		–		–		–		12	1277 1278
Ti + Al + ΣXi, Ti-6Al-4V	1	1074	4	598	7	1484 1488 1492 1493	7	1498	7	1501	–		10	326	–		12	1278 1279
Ti + Al + ΣXi, Ti-6Al-4V-3Co	–		–		–		–		–		–		–		–		12	1279
Ti + Cr + ΣXi	1	1077	4	601	–		–		–		–		–		–		12	1285
Ti + Cr + ΣXi, Ti-150A	1	1078 1089	–		–		–		–		–		–		–		12	1287
Ti + Cr + ΣXi, Ti-3Al-5Cr	–		–		–		–		–		–		–		–		12	1287
Ti + Fe + ΣXi	1	1080	4	604	–		–		–		–		–		–		–	
Ti + Fe + ΣXi, Ti-140A	1	1081	–		–		–		–		–		–		–		–	
Ti + Fe + ΣXi, Ti-2Cr-2Fe-2Mo	1	1081	–		–		–		–		–		–		–		–	
Ti + Fe + ΣXi, Russian, ferrocarbontitanium	1	1081	–		–		–		–		–		–		–		–	
Ti + Mn + ΣXi	1	1083	–		7	1503 1505	–		–		–		–		–		–	
Ti + Mn + ΣXi, RC-1308	1	1084	–		–		–		–		–		–		–		–	
Ti + Mn + ΣXi, RS-120	–		–		7	1503 1506	–		–		–		–		–		–	
Ti + Sn + ΣXi	–		–		–		–		–		–		–		–		12	1290
Ti + Sn + ΣXi, Ti-679	–		–		–		–		–		–		–		–		12	1293
Ti + V + ΣXi	1	1086	4	607	–		7	1508	–		–		–		–		12	1291
Ti + V + ΣXi, anodized	–		–		–		9	1293	–		–		–		–		–	
Ti + V + ΣXi, Ti-2.5Al-16V	1	1087	4	607	–		–		–		–		–		–		12	1294 1295
Ti + V + ΣXi, Ti-3Al-11Cr-13V	1	1087	–		–		–		–		–		–		–		–	
Ti + V + ΣXi, Ti-13V-11Cr-3Al	1	1087	4	607	–		–		–		–		–		–		–	

Substance Name	Thermal Conductivity		Specif. Heat		Thermal Radiative Properties								Thermal Diffusivity		Viscosity		Thermal Expansion	
					Emissivity		Reflectivity		Absorptivity		Transmissiv.							
	V.	Page	V.	Page	V.	Page	V.	Page	V.	Page	V.	Page	V.	Page	V.	Page	V.	Page
Titanium alloys: (continued)																		
Ti + V + ΣXi, MSM-2.5Al-16V		–	4	607		–		–		–		–		–		–		–
Ti + V + ΣXi, MST-2.5Al-16V		–	4	607		–		–		–		–		–		–		–
Ti + V + ΣXi, 120 VCA	1	1087		–		–		–		–		–		–		–		–
Ti + V + ΣXi, B-120VAC, anodized		–		–		–	9	1293		–		–		–		–		–
Ti + V + ΣXi, B-120VCA		–		–		–		–		–		–		–		–	12	1294
Titanium-aluminum intermetallic compound, TiAl		–		–	8	1338 1339	8	1341		–		–		–		–		–
Titanium beryllide, TiBe$_{12}$		–	5	328		–		–		–		–		–		–		–
Titanium boride, TiB$_2$	1	1358	5	378	8	703 705 707	8	710		–		–	10	484		–	13	778
Titanium boride + nickel powder		–		–		–	8	1435		–		–		–		–		–
Titanium boride + titanium, cermet		–		–	8	1410	8	1437		–		–		–		–		–
Titanium boride + titanium oxide powder		–		–	8	1489 1490	8	1493		–		–		–		–		–
Titanium boride + zirconium silicide powder		–		–		–	8	1495		–		–		–		–		–
Titanium bromides:																		
TiBr$_3$		–	5	778		–		–		–		–		–		–		–
TiBr$_4$		–	5	781		–		–		–		–		–		–		–
Titanium carbide, TiC	2	594	5	457	8	819 821 823 825		–		–		–	10	486		–	13	891
Titanium carbide + cobalt, cermet	2	725		–		–		–		–		–		–		–	13	1341
Titanium carbide + cobalt + niobium carbide, cermet	2	726		–		–		–		–		–		–		–		–
Titanium carbide + cobalt, Kennametal K138A		–		–		–		–		–		–		–		–	13	1341
Titanium carbide + nickel, Kennametal K150A		–		–	8	1413 1415		–		–		–		–		–		–
Titanium carbide + nickel, Kennametal K151A		–		–	8	1413 1415		–		–		–		–		–		–
Titanium carbide + nickel, Kennametal K152B		–		–	8	1413 1415		–		–		–		–		–		–
Titanium carbide + nickel, Kennametal K153B		–		–	8	1413 1415		–		–		–		–		–		–
Titanium carbide + nickel, Kennametal K161B	2	728		–		–		–		–		–		–		–		–
Titanium carbide + nickel, Kennametal K163B		–		–	8	1413 1415		–		–		–		–		–		–
Titanium carbide + nickel, Kennametal K184B		–		–	8	1413 1415		–		–		–		–		–		–
Titanium carbide + molybdenum + nickel + niobium carbide, cermet	2	727		–		–		–		–		–		–		–		–
Titanium carbide + nickel + ΣXi, cermet		–		–	8	1412 1415		–		–		–		–		–	13	1319

Substance Name	Thermal Conduc- tivity		Specif. Heat		Thermal Radiative Properties										Thermal Diffu- sivity		Visco- sity		Thermal Expan- sion	
					Emis- sivity		Reflec- tivity		Absorp- tivity		Trans- missiv.									
	V.	Page	V.	Page	V.	Page	V.	Page	V.	Page	V.	Page	V.	Page	V.	Page	V.	Page	V.	Page
Titanium carbide + nickel + niobium carbide, cermet	2	730		–		–		–		–		–		–		–		–		–
Titanium carbide + steel, T-420-G, cermet		–		–		–		–		–		–		–		–		13	1341
Titanium carbide + steel, T-520-G, cermet		–		–		–		–		–		–		–		–		13	1341
Titanium chlorides:																				
TiCl₃		–	5	893		–		–		–		–		–		–		–
TiCl₄		–	6s	91		–		–		–		–		–		–		–
Titanium-chromium intermetallic compound, TiCr₂		–		–	8	1343 1344	8	1346		–		–		–		–		–
Titanium dideuteride, TiD₂		–		–		–		–		–		–		–		–		13	1085
Titannium hydrides:																				
TiH		–		–		–		–		–		–		–		–		13	1088
TiH₂		–	5	1047		–		–		–		–		–		–		13	1085
Nonstoichiometric		–	5	1044		–		–		–		–		–		–			–
Titanium iodide, TiI₄	1	411	5	510		–		–		–		–		–		–		13	1122
Titanium-nickel intermetallic compound, TiNi	1	1361		–		–		–		–		–		–		–			–
Titanium nitride, TiN	2	668	5	1093	8	1084 1086		–		–		–	10	492		–		13	1147
Titanium oxides:																				
Rutile	2	203		–		–		–		–		–		–		–			–
TiO		–	5	243		–	8	454		–		–		–		–		13	390
TiO₂	2	202	5	246	8	456 458	8	461 473	8	475	8	476		–		–		13	392
Ti₂O₃		–	5	250		–		–		–		–		–		–		13	398
Ti₃O₅		–	5	253		–		–		–		–		–		–			–
Titania, dense	2	204		–		–		–		–		–		–		–			–
Titanium oxide + titanium, cermet		–		–		–	8	1439		–		–		–		–		13	1341
Titanium oxide + titanium chromium compound, cermet		–		–	8	1419 1420	8	1421		–		–		–		–			–
Titanium oxide + zirconium silicide, mixture		–		–		–	8	1507		–		–		–		–			–
Titanium phosphide, TiP		–		–	8	1105		–		–		–		–		–			–
Titanium silicide powder		–		–	8	1498 1500	8	1504		–		–		–		–			–
Titanium silicides:																				
TiSi		–	5	601		–		–		–		–		–		–		13	1213
TiSi₂		–	5	604	8	1163 1165 1167	8	1169		–		–		–		–		13	1213
Ti₅Si₃		–	5	607		–	8	1169		–		–		–		–		13	1213
Titanium tetrafluoride, TiF₄		–	5	1012		–		–		–		–		–		–			–
Toluene	3	242	6	285		–		–		–		–		–	11	218		–
m-Toluic acid		–	6s	91		–		–		–		–		–		–			–
o-Toluic acid		–	6s	91		–		–		–		–		–		–			–
p-Toluic acid		–	6s	91		–		–		–		–		–		–			–

Substance Name	Thermal Conductivity		Specif. Heat		Thermal Radiative Properties								Thermal Diffusivity		Viscosity		Thermal Expansion	
					Emissivity		Reflectivity		Absorptivity		Transmissiv.							
	V.	Page	V.	Page	V.	Page	V.	Page	V.	Page	V.	Page	V.	Page	V.	Page	V.	Page
m-Toluylic acid		–	6s	91	–		–		–		–		–		–		–	
o-Toluylic acid		–	6s	91	–		–		–		–		–		–		–	
p-Toluylic acid		–	6s	91	–		–		–		–		–		–		–	
Tomato		–		–	–		–		–		–		10	647	–		–	
Topaz	2	251		–	–		–		–		–		–		–		–	
Torr seal		–		–	–		–		–		–		–		–		13	1504
Tourmaline	2	855		–	–		–		–		–		–		–		–	
Tourmaline, brazil	2	855		–	–		–		–		–		–		–		–	
Transite	2	1107		–	–		–		–		–		10	568	–		–	
Tribromoborane		–	6s	2	–		–		–		–		–		–		–	
Tribromochloromethane		–	6s	22	–		–		–		–		–		–		–	
Tribromofluoromethane		–	6s	91	–		–		–		–		–		–		–	
Tribromoethane		–	6s	5	–		–		–		–		–		–		–	
Tribromohydrin		–	6s	91	–		–		–		–		–		–		–	
1,2,3-Tribromopropane		–	6s	91	–		–		–		–		–		–		–	
Trichloroborane		–	6s	2	–		–		–		–		–		–		–	
1,1,1-Trichloroethane		–	6s	91	–		–		–		–		–		–		–	
Trichloroethylene		–	6s	91	–		–		–		–		–		–		–	
Trichlorofluoromethane	3	183	6	200	–		–		–		–		–		11	220	–	
Trichlorohydrin		–	6s	91	–		–		–		–		–		–		–	
Trichloromethane	3	161	6	166	–		–		–		–		–		–		–	
Trichlorophosphine		–	6s	75	–		–		–		–		–		–		–	
1,2,3-Trichloropropane		–	6s	91	–		–		–		–		–		–		–	
Trichlorosilane		–	6s	92	–		–		–		–		–		–		–	
Trichlorotrifluoroethane	3	201	6	224	–		–		–		–		–		11	225	–	
1,1,1-Trichloro-2,2,2-trifluoroethane		–	6s	92	–		–		–		–		–		–		–	
Tridecane		–	6s	92	–		–		–		–		–		–		–	
Trideuteriomethane		–	6s	59	–		–		–		–		–		–		–	
Trideuteriomonotritiomethane		–	6s	59	–		–		–		–		–		–		–	
Trideuteriotritiomethane		–	6s	59	–		–		–		–		–		–		–	
Tridymite		–	5	213	–		–		–		–		–		–		13	353
Triethylmethylmethane		–	6s	38	–		–		–		–		–		–		–	
Trifluoroborane	3	99		–	–		–		–		–		–		–		–	
1,1,1-Trifluoroethane		–	6s	92	–		–		–		–		–		–		–	
Trifluoromethane				–	–		–		–		–		–		11	230	–	
Trifluoroiodomethane		–	6s	92	–		–		–		–		–		–		–	
Trifluorophosphine		–	6s	75	–		–		–		–		–		–		–	
Trifluorotrichloroethane	3	201		–	–		–		–		–		–		–		–	
Trimethylamine		–	6s	93	–		–		–		–		–		–		–	
sym-Trimethylbenzene		–	6s	57	–		–		–		–		–		–		–	
unsym-Trimethylbenzene		–	6s	93	–		–		–		–		–		–		–	
1,2,4-Trimethylbenzene		–	6s	93	–		–		–		–		–		–		–	

Substance Name	Thermal Conductivity		Specif. Heat		Thermal Radiative Properties								Thermal Diffusivity		Viscosity		Thermal Expansion	
					Emissivity		Reflectivity		Absorptivity		Transmissiv.							
	V.	Page	V.	Page	V.	Page	V.	Page	V.	Page	V.	Page	V.	Page	V.	Page	V.	Page
1,3,5-Trimethylbenzene		–	6s	57		–		–		–		–		–		–		–
2,2,3-Trimethylbutane		–	6s	93		–		–		–		–		–		–		–
Trimethyl carbinol		–	6s	67		–		–		–		–		–		–		–
Trimethylene		–	6s	26		–		–		–		–		–		–		–
Trimethylene bromide		–	6s	27		–		–		–		–		–		–		–
Trimethylene dibromide		–	6s	27		–		–		–		–		–		–		–
Trimethylethylene		–	6s	61		–		–		–		–		–		–		–
2,2,4-Trimethylpentane		–	6s	94		–		–		–		–		–		–		–
2,3,3-Trimethylpentane		–	6s	94		–		–		–		–		–		–		–
2,3,4-Trimethylpentane		–	6s	94		–		–		–		–		–		–		–
2,4,4-Trimethyl-2-pentene		–	6s	94		–		–		–		–		–		–		–
Trinitrotoluene	2	1007		–		–		–		–		–		–		–		–
Triptane		–	6s	93		–		–		–		–		–		–		–
Tritimethane		–	6s	58		–		–		–		–		–		–		–
Tritiomethane		–	6s	58		–		–		–		–		–		–		–
Tritium	3	87		–		–		–		–		–		–		–		–
Tritium sulfide		–	6s	51		–		–		–		–		–		–		–
Tritritiomethane		–	6s	59		–		–		–		–		–		–		–
Triuranium disilicide + triuranium monosilicide, mixture		–	5	622		–		–		–		–		–		–		–
Trizinc diiodine hexasulfide		–		–		–		–		–	8	1236		–		–		–
Tuballoy	1	429		–		–		–		–		–		–		–		–
Tuff	2	856		–		–		–		–		–		–		–		–
Tungstates, miscellaneous		–		–		–	8	665		–		–		–		–		–
Tungsten, W	1	415	4	263	7	776 782 790 796 808 810	7	812 814 819 823	7	825		–	10	198		–	12	354
Tungsten alloys:																		
W + Co		–	4	459		–		...		–		–		–		–	12	757
W + Cu		–		–		–		–		–		–		–		–	12	789
W + Fe		–	4	462		–		–		–		–		–		–		–
W + Mo		–		–	7	1043 1045		–		–		–		–		–		–
W + Re	1	855		–	7	1048 1051		–		–		–		–		–	12	973
W + Re, VR-27-VT		–		–	7	1049		–		–		...		–		–		–
W + Ta		–		–		–		–		–		–		–		–	12	981
W + Ti		–		–		–		–		–		–		–		–	12	988
W + Fe + ΣXi	1	1090		–		–		–		...		–		–		...		–
W + Fe + ΣXi, Russian, ferrotungsten	1	1090		–		–		–		...		–		–		–		–
W + Ni + ΣXi	1	1091		–		–		–		–		–		–		–	12	1297
W + Ni + ΣXi, mallory 1000		–		–		–		–		–		–		–		–	12	1297 1299
W + Re + ΣXi		–		–		–		–		–		–		–		–	12	1300

Substance Name	Thermal Conductivity		Specif. Heat		Thermal Radiative Properties										Thermal Diffusivity		Viscosity		Thermal Expansion	
					Emissivity		Reflectivity		Absorptivity		Transmissiv.									
	V.	Page	V.	Page	V.	Page	V.	Page	V.	Page	V.	Page	V.	Page	V.	Page	V.	Page
Tungsten arsenide, W_3As_7	1	1364		–		–		–		–		–			–			–			–
Tungsten borides:																				
WB	1	1365	5	382	8	730	8	735		–		–			–			–	13	798
WB_2		–		–		–		–		–		–	10	495			–			–
W_2B		–	5	385		–		–		–		–			–			–			–
W_2B_5		–	5	388	8	732		–		–		–			–			–			–
Tungsten carbide + cobalt, cermet		–	5	1282		–		–		–		–			–			–	13	1322
Tungsten carbides:																				
WC	2	598	5	460	8	827 829 832		–		–		–			–			–	13	898
W_2C		–		–	8	829 832		–		–		–			–			–			–
Tungsten oxide + zinc oxide, mixture	2	422		–		–		–		–		–			–			–			–
Tungsten oxides:																				
WO_2		–		–		–		–		–		–			–			–	13	402
WO_3	2	209	5	256		–		–		–		–			–			–	13	405
Tungsten selenide, WSe_2	1	1368		–		–		–		–		–			–			–			–
Tungsten silicides:																				
WSi_2	1	1369	5	610		–	8	1176		–		–			–			–	13	1205
W_3Si		–		–	8	1173		–		–		–			–			–			–
Tungsten telluride, WTe_2	1	1370		–		–		–		–		–			–			–			–
Tylose gel		–		–		–		–		–		–	10	648			–			–
Undecane		–	6s	95		–		–		–		–			–			–			–
Unitane		–		–		–	8	465		–	8	475			–			–			–
Unitemp No. 41		–		–	7	936 1352		–		–		–			–			–			–
Uranic chloride		–	5	899		–		–		–		–			–			–			–
Uranic iodide		–	5	513		–		–		–		–			–			–			–
Uranic oxide, UO_2	2	210	5	259		–		–		–		–			–			–			–
Uranium, U	1	429	4	268	7	828 834 838		–		–		–	10	205			–	12	365
Uranium, X–metal	1	429	4	268		–		–		–		–			–			–			–
Uranium alloys:																				
U + Al	1	858		–		–		–		–		–			–			–			–
U + Cr	1	859		–		–		–		–		–			–			–	12	728
U + Fe	1	862		–		–		–		–		–			–			–			–
U + Mg	1	863		–		–		–		–		–			–			–	12	889
U + Mo	1	864		–		–		–		–		–			–			–	12	919
U + Nb	1	867		–	7	1053 1056		–		–		–			–			–			–
U + Si	1	868		–		–		–		–		–			–			–			–
U + Zr	1	871		–		–		–		–		–			–			–	12	997

Substance Name	Thermal Conductivity		Specif. Heat		Thermal Radiative Properties								Thermal Diffusivity		Viscosity		Thermal Expansion	
					Emissivity		Reflectivity		Absorptivity		Transmissiv.							
	V.	Page	V.	Page	V.	Page	V.	Page	V.	Page	V.	Page	V.	Page	V.	Page	V.	Page
Uranium alloys: (continued)																		
U + Zr + ΣXi	1	1097		–		–		–		–		–		–		–		–
U + Al + ΣXi		–		–		–		–		–		–		–		–	12	1301
U + Mo + ΣXi	1	1094		–		–		–		–		–		–		–	12	1303
U + Mo + ΣXi, Fissium	1	1095		–		–		–		–		–		–		–		–
U + Mo + ΣXi, Uranium-3 percent fissium alloy	1	1095		–		–		–		–		–		–		–		–
U + Mo + ΣXi, Uranium-5 percent fissium alloy	1	1095 1097		–		–		–		–		–		–		–		–
U + Mo + ΣXi, Uranium-8 percent fissium alloy	1	1095		–		–		–		–		–		–		–		–
U + Mo + ΣXi, Uranium-10 percent fissium alloy	1	1095		–		–		–		–		–		–		–		–
U + Nb + ΣXi		–		–		–		–		–		–		–		–	12	1304
U + Pu + ΣXi		–		–		–		–		–		–		–		–	12	1306
U + V + ΣXi		–		–		–		–		–		–		–		–	12	1309
U + Zr + ΣXi		–		–	7	1511 1514		–		–		–		–		–		–
Uranium borides:																		
UB$_2$		–		–		–		–		–		–		–		–	13	798
UB$_{12}$		–		–	8	732		–		–		–		–		–		–
Uranium carbides:																		
UC	2	601	5	463		–		–		–		–	10	496		–	13	902
UC$_2$	2	605	5	466	8	848 852		–		–		–		–		–	13	911
U$_2$C$_3$		–	5	472		–		–		–		–		–		–	13	913
Nonstoichiometric		–	5	469		–		–		–		–		–		–		–
Uranium carbide + uranium, cermet	2	731		–		–		–		–		–		–		–	13	1342
Uranium carbide + zirconium carbide, mixture		–		–		–		–		–		–		–		–	13	943
Uranium chlorides:																		
UCl$_3$		–	5	896		–		–		–		–		–		–		–
UCl$_4$		–	5	899		–		–		–		–		–		–		–
Uranium dioxide + chromium, cermet	2	732		–		–		–		–		–		–		–	13	1342
Uranium dioxide + molybdenum, cermet	2	735		–		–		–		–		–		–		–	13	1326
Uranium dioxide + niobium, cermet	2	738		–		–		–		–		–		–		–		–
Uranium dioxide + stainless steel, cermet	2	741		–		–		–		–		–		–		–	13	1330
Uranium dioxide + uranium, cermet	2 1	744 1442		–		–		–		–		–		–		–		–
Uranium dioxide + uranium oxycarbide, mixture		–		–		–		–		–		–	10	447		–		–
Uranium dioxide + yttrium oxide, mixture	2	428 432		–		–		–		–		–		–		–		–
Uranium dioxide + zirconium, cermet	2	746		–		–		–		–		–		–		–	13	1342
Uranium dioxide + zirconium dioxide, mixture	2	429		–		–		–		–		–		–		–		–

Substance Name	Thermal Conductivity V.	Page	Specif. Heat V.	Page	Thermal Radiative Properties Emissivity V.	Page	Reflectivity V.	Page	Absorptivity V.	Page	Transmissiv. V.	Page	Thermal Diffusivity V.	Page	Viscosity V.	Page	Thermal Expansion V.	Page
Uranium fluorides:																		
UF$_4$		–	5	1015		–		–		–		–		–		–		–
UF$_6$		–	5	1018		–		–		–		–		–		–		–
Uranium hydride, UH$_3$		–	5	1050		–		–		–		–		–		–		–
Uranium iodide, UI$_4$		–	5	513		–		–		–		–		–		–		–
Uranium nitride, UN	2	672	5	1096		–		–		–		–	10	500		–	13	1152
Uranium nitride, nonstoichiometric		–	5	1099		–		–		–		–		–		–		–
Uranium oxides:																		
UO$_2$	2	210	5	259	8	478 485		–		–	8	486	10	402		–	13	413
UO$_2$, powder	2	1040		–		–		–		–		–		–		–		–
UO$_3$		–	5	262		–		–		–		–		–		–		–
U$_3$O$_8$	2	237	5	265		–		–		–		–		–		–	13	421
U$_4$O$_9$		–	5	269		–		–		–		–		–		–	13	426
Uranium oxycarbide		–		–		–		–		–		–	10	502		–		–
Uranium phosphide, UP		–		–		–		–		–		–	10	505		–	13	1178
Uranium plutonium carbide, UPuC		–		–		–		–		–		–		–		–	13	917
Uranium selenide, USe		–		–		–		–		–		–		–		–	13	1192
Uranium silicides:																		
USi$_2$		–	5	619		–		–		–		–		–		–		–
βUSi$_2$		–		–		–		–		–		–		–		–	13	1213
USi$_3$		–	5	616		–		–		–		–		–		–		–
U$_3$Si		–	5	613		–		–		–		–		–		–	13	1213
U$_3$Si$_2$		–		–		–		–		–		–		–		–	13	1213
Uranium sulfide, US		–		–		–		–		–		–	10	507		–	13	1229
Uranium trisilicide + tungsten, cermet		–		–		–		–		–		–		–		–	13	1342
Uranus 10		–		–		–	7	1283		–		–		–		–		–
Uranyl oxide		–	5	262		–		–		–		–		–		–		–
Uranyl uranate		–	5	265		–		–		–		–		–		–		–
Valerianic ether		–	6s	38		–		–		–		–		–		–		–
Valerylene		–	6s	73		–		–		–		–		–		–		–
Vanadic anhydride		–	5	281		–		–		–		–		–		–		–
Vanadium, V	1	441	4	271	7	840	7	844 848	7	850		–	10	209		–	12	373
Vanadium alloys:																		
V + Al		–	4	465		–		–		–		–		–		–		–
V + Cr		–		–		–		–		–		–		–		–	12	729
V + Fe	1	874	4	471		–		–		–		–		–		–		–
V + Fe, Russian, ferrovanadium	1	875		–		–		–		–		–		–		–		–
V + Mo		–		–		–		–		–		–		–		–	12	923
V + Nb		–		–		–		–		–		–		–		–	12	955
V + Sb		–	4	468		–		–		–		–		–		–		–
V + Sn		–	4	474		–		–		–		–		–		–		–

Substance Name	Thermal Conductivity		Specif. Heat		Thermal Radiative Properties								Thermal Diffusivity		Viscosity		Thermal Expansion	
					Emissivity		Reflectivity		Absorptivity		Transmissiv.							
	V.	Page	V.	Page	V.	Page	V.	Page	V.	Page	V.	Page	V.	Page	V.	Page	V.	Page
Vanadium alloys: (continued)																		
V + Ti		–	4	477		–		–		–		–		–		–	12	989
V + Y	1	877		–		–		–		–		–		–		–		–
V + Cr + ΣXi		–		–		–		–		–		–		–		–	12	1310
V + Ti + ΣXi		–		–		–		–		–		–		–		–	12	1311
Vanadium boride, VB$_2$		–		–	8	732		–		–		–		–		–	13	783
Vanadium carbides:																		
VC	2	606	5	475	8	850		–		–		–		–		–	13	922
V$_2$C		–		–		–		–		–		–		–		–	13	596
Vanadium chlorides:																		
VCl$_2$		–	5	902		–		–		–		–		–		–		–
VCl$_3$		–	5	905		–		–		–		–		–		–		–
Vanadium fluoride, VF$_3$		–	5	1021		–		–		–		–		–		–		–
Vanadium hydrides:																		
VH		–		–		–		–		–		–		–		–	13	1089
Nonstoichiometric		–	5	1053		–		–		–		–		–		–		–
Vanadium nitrides:																		
VN		–	5	1103	8	1087		–		–		–		–		–	13	1163
V$_3$N		–		–	8	1087		–		–		–		–		–		–
Vanadium oxides:																		
VO		–	5	272		–		–		–		–		–		–		–
VO$_2$		–		–		–		–		–		–		–		–	13	430
V$_2$O$_3$		–	5	275		–		–		–		–		–		–		–
V$_2$O$_4$		–	5	278		–		–		–		–		–		–		–
V$_2$O$_5$		–	5	281		–	8	546		–		–		–		–	13	432
Vanadium phosphate, VPO$_5$		–		–		–		–		–		–		–		–	13	691
Vanadium silicides:																		
VSi$_2$		–	5	628	8	1173		–		–		–		–		–	13	1214
V$_3$Si		–	5	625		–		–		–		–		–		–	13	1214
V$_5$Si$_3$		–	5	631		–		–		–		–		–		–	13	1214
Verilite		–		–		–		–		–		–		–		–	12	1040
Vermiculite mica, granulated	2	825		–		–		–		–		–		–		–		–
Vinegar acid		–	6s	1		–		–		–		–		–		–		–
Vinegar naphtha		–	6s	35		–		–		–		–		–		–		–
Vinyl acetate		–	6s	95		–		–		–		–		–		–		–
Vinylbenzene		–	6s	84		–		–		–		–		–		–		–
Vinylethylene		–	6s	5		–		–		–		–		–		–		–
Vinyl fluoride		–	6s	41		–		–		–		–		–		–		–
Vinylidene fluoride		–	6s	30		–		–		–		–		–		–		–
Vitreosil, I.R.		–		–		–		–		–		–		–		–	13	365
Vitreous silica	2	184 185 187		–	8	403 405	8	409 413	8	415	8	417 426		–		–	13	360 362 365

Substance Name	Thermal Conductivity		Specif. Heat		Thermal Radiative Properties								Thermal Diffusivity		Viscosity		Thermal Expansion	
					Emissivity		Reflectivity		Absorptivity		Transmissiv.							
	V.	Page	V.	Page	V.	Page	V.	Page	V.	Page	V.	Page	V.	Page	V.	Page	V.	Page
Wallboard	2	1131		–		–		–		–		–		–		–		–
Wallboard, cornstalk	2	1111		–		–		–		–		–		–		–		–
Walnut	2	1089		–		–		–		–		–		–		–		–
Water	3	120	6	102		–		–		–		–	10	390	11	94	13	261
Water, dideuterated		–	6s	95		–		–		–		–		–		–		–
Whiting		–		–		–		–		–		–	10	414		–		–
Wolfram, tungsten	1	415	4	263		–		–		–		–		–		–		–
Wollastonite	2	859		–		–	8	618		–		–		–		–	13	705
Wood, american white	2	1090		–		–		–		–		–		–		–		–
Wood, box	2	1061		–		–		–		–		–		–		–		–
Wood, felt	2	1133		–		–		–		–		–		–		–		–
Wood, fiber blanket	2	1132		–		–		–		–		–		–		–		–
Wood, greenheart	2	1074		–		–		–		–		–		–		–		–
Wood, hardwood	2	1075		–		–		–		–		–		–		–		–
Wood, mineral board processed	2	1141		–		–		–		–		–		–		–		–
Wood, redwood	2	1084		–		–		–		–		–		–		–		–
Wood, redwood bark	2	1084		–		–		–		–		–		–		–		–
Wood, sawdust	2	1085		–		–		–		–		–		–		–		–
Wood, white	2	1090		–		–		–		–		–		–		–		–
Wool, angora	2	1092		–		–		–		–		–		–		–		–
Wool, mineral	2	1147		–		–		–		–		–		–		–		–
Wool, sheep	2	1092		–		–		–		–		–		–		–		–
Wulfenite		–		–		–	8	1673		–		–		–		–		–
Wustite		–		–		–		–		–		–		–		–	13	273
Xenon, Xe	3	88	6	57		–		–		–		–		–	11	62	13	170
Xenon tetrafluoride, XeF_4		–	5	1024		–		–		–		–		–		–		–
m-Xylene		–	6s	98		–		–		–		–		–		–		–
o-Xylene		–	6s	99		–		–		–		–		–		–		–
p-Xylene		–	6s	101		–		–		–		–		–		–		–
Ytterbia		–	5	284		–		–		–		–		–		–		–
Ytterbium, Yb	1	446	4	274		–		–		–		–	10	212		–	12	382
Ytterbium borides:																		
YbB$_6$		–		–	8	723		–		–		–		–		–		–
YbB$_{12}$		–		–		–		–		–		–		–		–	13	791
Ytterbium gallium oxide, $Yb_3Ga_5O_{12}$		–	5	1620		–		–		–		–		–		–		–
Ytterbium hydride, YbH_2		–		–		–		–		–		–		–		–	13	1091
Ytterbium oxide, Yb_2O_3		–	5	284	8	488		–		–	8	490		–		–	13	435
Ytterbium-zinc intermetallic compound, $YbZn_2$		–		–		–		–		–		–		–		–	12	625
Yttria	2	240	5	287		–		–		–		–		–		–		–
Yttrium, Y	1	449	4	278	7	853		–		–		–	10	213		–	12	387
Yttrium aluminate, $Y_3Al_5O_{12}$	2	308		–		–		–		–		–		–		–		–
Yttrium aluminate garnet		–		–		–	8	579		–		–		–		–		–

Substance Name	Thermal Conductivity		Specif. Heat		Thermal Radiative Properties								Thermal Diffusivity		Viscosity		Thermal Expansion	
					Emissivity		Reflectivity		Absorptivity		Transmissiv.							
	V.	Page	V.	Page	V.	Page	V.	Page	V.	Page	V.	Page	V.	Page	V.	Page	V.	Page
Yttrium aluminate garnet, YAG		–		–		–	8	579		–		–		–		–		–
Yttrium borides:																		
YB_6		–		–	8	723		–		–		–		–		–		–
YB_{12}		–		–		–		–		–		–		–		–	13	791
Yttrium carbides:																		
YC		–		–	8	852		–		–		–		–		–		–
YC_2		–		–	8	852		–		–		–		–		–	13	937
Y_2C_3		–		–	8	852		–		–		–		–		–	13	937
Yttrium deuterides:																		
YD_2		–	5	1062		–		–		–		–		–		–		–
YD_3		–	5	1066		–		–		–		–		–		–		–
Yttrium gallium oxide, $Y_3Ga_5O_{12}$		–	5	1623		–		–		–		–		–		–		–
Yttrium hydrides:																		
YH_2		–	5	1056		–		–		–		–		–		–		–
YH_3		–	5	1059		–		–		–		–		–		–		–
Yttrium iron garnet	2	311		–		–		–		–		–		–		–		–
Yttrium nitride		–		–		–		–		–	8	1090		–		–		–
Yttrium oxide, Y_2O_3	2	240	5	287	8	492 494 496 498	8	501		–	8	504	10	407		–	13	438
Yttrium oxide + zirconium oxide, mixture	2	449		–		–		–		–		–	10	454		–		–
Yttrium oxide + zirconium oxide + ΣXi, mixture	2	537		–		–		–		–		–		–		–		–
Yttrium phosphate		–		–		–	8	607		–		–		–		–		–
Yttrium silicate, Y_2SiO_5		–		–		–		–		–		–		–		–	13	728
Yttrium vanadate		–		–	8	669	8	671		–		–		–		–		–
Zinc, Zn	1	453	4	281	7	855 857	7	860	7	864 866 869		–	10	216		–	12	391
Zinc alloys:																		
Zn + Al	1	880		–		–	7	1059		–		–		–		–	12	665
Zn + Al, Zamak Nr 400	1	880		–		–		–		–		–		–		–		–
Zn + Al, Zamak Nr 410	1	1098		–		–		–		–		–		–		–		–
Zn + Al, Zamak Nr 430	1	1098		–		–		–		–		–		–		–		–
Zn + Cd	1	881		–		–		–		–		–		–		–		–
Zn + Cu		–	4	480		–		–		–		–		–		–	12	797
Zn + Mg		–	4	483		–		–		–		–		–		–		–
Zn + Zr		–	4	486		–		–		–		–		–		–		–
Zn + Al + ΣXi	1	1098		–		–		–		–		–		–		–		–
Zn + Pb + ΣXi	1	1099		–		–		–		–		–		–		–		–
Zinc aluminum oxide, $ZnO \cdot Al_2O_3$		–		–		–		–		–		–		–		–	13	485
Zinc antimonide		–		–		–	8	1348		–		–		–		–		–
Zinc arsenide, Zn_3As_2		–		–		–		–		–		–		–		–	13	752
Zinc chloride, $ZnCl_2$	2	626	5	908		–		–		–		–		–		–		–

| Substance Name | Thermal Conductivity | | Specif. Heat | | Thermal Radiative Properties | | | | | | | | Thermal Diffusivity | | Viscosity | | Thermal Expansion | |
| | | | | | Emissivity | | Reflectivity | | Absorptivity | | Transmissiv. | | | | | | | |
	V.	Page	V.	Page	V.	Page	V.	Page	V.	Page	V.	Page	V.	Page	V.	Page	V.	Page
Zinc fluoride, ZnF_2		–	5	1027		–		–		–		–		–		–	13	1071
Zinc germanium oxide, $2ZnO \cdot GeO_2$		–		–		–		–		–		–		–		–	13	500
Zinc germanium phosphide, $ZnGeP_2$	2	792		–		–		–		–		–		–		–		–
Zinc iron oxide, $ZnO \cdot Fe_2O_3$	2	314	5	1626		–		–		–		–		–		–		–
Zinc diiodine tetrasulfide		–		–		–		–		–	8	1236		–		–		–
Zinc ferrite		–	5	1626		–		–		–		–		–		–		–
Zinc oxide, AZO-33, powder compact		–		–		–	8	510 518		–		–		–		–		–
Zinc oxide, AZO-55LO, powder compact		–		–		–	8	510 518		–		–		–		–		–
Zinc oxide, AZO-66, powder compact		–		–		–	8	510 518		–		–		–		–		–
Zinc oxide, Kodak 515		–		–		–		–	8	521		–		–		–		–
Zinc oxide, ZnO	2	243	5	290	8	506	8	507	8	519 521	8	522	10	408		–	13	444
Zinc phosphates:																		
$Zn(PO_3)_2$		–		–		–		–		–		–		–		–	13	684
$Zn_2P_2O_7$		–		–		–		–		–		–		–		–	13	691
$Zn_3(PO_4)_2$		–		–		–		–		–		–		–		–	13	681
Zinc selenide, ZnSe	1	1371		–	8	1112 1114	8	1117	8	1121	8	1123		–		–	13	1187
Zinc silicon arsenide, $ZnSAs_2$	1	1374		–		–		–		–		–		–		–		–
Zinc silicate, Zn_2SiO_4		–	5	1629		–		–		–		–		–		–	13	722
Zinc sulfate heptahydrate, $ZnSO_4 \cdot 7H_2O$	2	694	5	1224		–		–		–		–		–		–		–
Zinc sulfide, ZnS		–	5	714	8	1213 1215	8	1217 1222	8	1224 1226	8	1227		–		–	13	1232
Zinc telluride, ZnTe		–		–		–	8	1266		–	8	1268		–		–	13	1265
Zinc titanium oxide, $2ZnO \cdot TiO_2$		–	5	1632		–	8	660	8	662		–		–		–		–
Zinc vanadium oxides:																		
$ZnO \cdot V_2O_5$		–		–		–		–		–		–		–		–	13	598
$2ZnO \cdot V_2O_5$		–		–		–		–		–		–		–		–	13	599
$3ZnO \cdot V_2O_5$		–		–		–		–		–		–		–		–	13	600
Zinc-zirconium intermetallic compound, Zn_2Zr		–		–		–		–		–		–		–		–	12	628
Zinc zirconium silicate		–		–		–		–	8	616		–		–		–		–
Zircoa, zirconium dioxide		–		–		–	8	540		–		–		–		–		–
Zircon, zirconium silicon tetraoxide		–	5	1635	8	1685 1687	8	613	8	615		–		–		–		–
Zircon, 475	2	318		–		–		–		–		–		–		–		–
Zircon, Brazil	2	318		–		–		–		–		–		–		–		–
Zircon, Florida		–		–		–		–		–		–		–		–	13	725
Zircon, Taylor		–		–		–		–		–		–		–		–	13	725
Zircon Tam	2	318		–		–		–		–		–		–		–		–
Zirconates		–		–		–	8	675		–		–		–		–		–
Zirconia	2	246	5	293	8	524 526 529 531 533	8	536	8	544		–		–		–		–

Substance Name	Thermal Conductivity V.	Page	Specif. Heat V.	Page	Thermal Radiative Properties — Emissivity V.	Page	Reflectivity V.	Page	Absorptivity V.	Page	Transmissiv. V.	Page	Thermal Diffusivity V.	Page	Viscosity V.	Page	Thermal Expansion V.	Page
Zirconia, TAM		-		-		-	8	540		-		-		-		-		-
Zirconia, stabilized	2	522		-		-		-		-		-		-		-		-
Zirconium, Zr	1	461	4	287	7	872 875 878	7	881	7	883		-	10	220		-	12	400
Zirconium, 2 percent impurities	1	1112		-		-		-		-		-		-		-		-
Zirconium alloys:																		
Zr + Ag		-	4	498		-		-		-		-		-		-		-
Zr + Al	1	882		-		-		-		-		-		-		-		-
Zr + Dy		-		-		-		-		-		-	10	262		-		-
Zr + Hf	1	883		-		-		-		-		-		-		-	12	823
Zr + In		-	4	489		-		-		-		-		-		-		-
Zr + Fe		-	4	492		-		-		-		-		-		-		-
Zr + Nb	1	886	4	495		-		-		-		-		-		-	12	959
Zr + Sn	1	887	4	501		-		-		-		-	10	264		-	12	986
Zr + Ti	1	890	4	504		-		-		-		-		-		-	12	993
Zr + U	1	891	4	507		-		-		-		-	10	267		-	12	998
Zr + Al + ΣXi	1	1100		-		-		-		-		-	10	329		-		-
Zr + Hf + ΣXi	1	1101	4	613	7	1517		-		-		-		-		-		-
Zr + Fe + ΣXi		-	4	610		-		-		-		-		-		-		-
Zr + Mo + ΣXi	1	1104		-		-		-		-		-		-		-		-
Zr + Sn + ΣXi	1	1108		-	7	1519 1522		-		-		-		-		-	12	1312
Zr + Sn + ΣXi, Zircaloy-2	1	888	4	501	7	1520 1523		-		-		-	10	265		-	12	987 1313
Zr + Sn + ΣXi, Zircaloy-4	1	888		-		-		-		-		-		-		-	12	1313 1314
Zr + Ta + ΣXi	1	1105		-		-		-		-		-		-		-		-
Zr + Ti + ΣXi		-		-		-		-		-		-		-		-	12	1317
Zr + U + ΣXi	1	1111	4	616	7	1525 1528		-		-		-		-		-		-
Zr + Ti + ΣXi, No. 7		-		-		-		-		-		-		-		-	12	1318
Zr + Ti + ΣXi, No. 9		-		-		-		-		-		-		-		-	12	1318
Zirconium beryllide		-	5	331		-		-		-		-		-		-		-
Zirconium 13-beryllide, $ZrBe_{13}$		-	5	331		-		-		-		-		-		-		-
Zirconium boride + chromium, cermet		-		-		-		-		-		-		-		-	13	1342
Zirconium borides:																		
ZrB	1	1375		-		-		-		-		-		-		-		-
ZrB_2		-	5	391	8	713 715 717	8	720		-		-	10	509		-	13	784
ZrB_{12}		-		-		-		-		-		-		-		-	13	793
Zirconium carbide, ZrC	2	609	5	478	8	833 835 837 841	8	843		-		-	10	511		-	13	926
Zirconium chromium oxide, $ZrO_2 \cdot Cr_2O_3$		-		-		-		-		-		-		-		-	13	490

Substance Name	Thermal Conductivity		Specif. Heat		Thermal Radiative Properties								Thermal Diffusivity		Viscosity		Thermal Expansion	
					Emissivity		Reflectivity		Absorptivity		Transmissiv.							
	V.	Page	V.	Page	V.	Page	V.	Page	V.	Page	V.	Page	V.	Page	V.	Page	V.	Page
Zirconium hydrides:																		
ZrH	2	793	–		–		–		–		–		–		–		–	
ZrH$_2$	–		5	1072	–		–		–		–		–		–		13	1093
Nonstoichiometric	–		5	1069	–		–		–		–		–		–		–	
Zirconium hydride + uranium, cermet	–						–		–		–		10	540	–		–	
Zirconium iodide	1	462 463	–		–		–		–		–		–		–		–	
Zirconium nitride, ZrN	2	675	5	1106	8	1078 1080 1082	–		–		–		10	514	–		13	1156
Zirconoum oxide, Norton RZ 5601	–		–		8	534					–				–		–	
Zirconium oxide, ZrO$_2$	2	246	5	293	8	524 526 529 531 533	8	536	8	544	–		10	409	–		13	451
Zirconium oxide, stabilized	–		–		8	527 532	8	538	8	544	–		–		–		–	
Zirconium oxide + aluminum, cermet	–		–		–		8	1442	–		–		–		–		–	
Zirconium oxide powder, SP 500	–		–		–		8	510 511 512	8	521	–		–		–		–	
Zirconium oxide + titanium, cermet	2	749	5	1285	–		–		–		–		–		–		13	1333
Zirconium oxide + yttrium oxide + zirconium, cermet	2	753	–		–		–		–		–		–		–		–	
Zirconium oxide + zirconium, cermet	2 1	752 1444	–		–		–		–		–		–		–		13	1337
Zirconium phosphates:																		
ZrPO$_4$	–		–		–		–		–		–		–		–		13	687
ZrP$_2$O$_7$	–		–		–		–		–		–		–		–		13	687
Zr$_2$P$_2$O$_9$	–		–		–		–		–		–		–		–		13	687
Zirconium silicates:																		
ZrSiO$_4$	2	317	5	1635	8	610	8	612	8	615	–		–		–		13	724
Natural	2	318	–		–		–		–		–		–		–		–	
Ultrox 1000W	–		–		–		8	613	–		–		–		–		–	
Zircon	2	318	–		–		–		–		–		–		–		–	
Zircon, Brazil	2	318	–		–		–		–		–		–		–		–	
Zircon Tam	2	318	–		–		–		–		–		–		–		–	
Zirconium silicide, ZrSi$_2$	–		–		8	1173	8	1176	–		–		–		–		13	1214
Zirconium tetrachloride, ZrCl$_4$	–		5	911	–		–		–		–		–		–		–	
Zirconium tetrafluoride, ZrF$_4$	–		5	1030	–		–		–		–		–		–		–	
Zirconium titanium oxide, ZrO$_2 \cdot$TiO$_2$	–		–		–		–		–		–		–		–		13	574
Zirconium tungsten oxide, ZrO$_2 \cdot$2WO$_3$	–		–		–		–		–		–		–		–		13	594
Zircopax	–		–		8	611	8	613	–		–		–		–		–